T0302190

Composites Innovation

Composites Innovation
Perspectives on Advancing the Industry

Edited by
Probir Guha

CRC Press
Taylor & Francis Group
Boca Raton London New York

CRC Press is an imprint of the
Taylor & Francis Group, an **informa** business

First edition published 2022
by CRC Press
6000 Broken Sound Parkway NW, Suite 300, Boca Raton, FL 33487-2742

and by CRC Press
2 Park Square, Milton Park, Abingdon, Oxon, OX14 4RN

© 2022 Taylor & Francis Group, LLC

CRC Press is an imprint of Taylor & Francis Group, LLC

Library of Congress Cataloging-in-Publication Data
Names: Guha, Probir, editor.
Title: Composites innovation : perspectives on advancing the industry /
edited by Probir Guha.
Description: First edition. | Boca Raton : CRC Press, [2022] | Includes
bibliographical references and index. | Summary: "This book provides a panoramic view of innovations in the composites industry. The book is arranged in five segments including: how composites fit into our world, the basics of the technology, customer insights, discussions from outside the world of composites, and paths forward to find competitive and effective solutions in a timely manner. The work is a resource for composites business leaders, researchers, and industry professionals to pioneer new solutions with composite materials"— Provided by publisher.
Identifiers: LCCN 2021031672 (print) | LCCN 2021031673 (ebook) |
ISBN 9780367752606 (hbk) | ISBN 9780367752613 (pbk) | ISBN 9781003161738 (ebk)
Subjects: LCSH: Composite materials—Technological innovations.
Classification: LCC TA418.9.C6 C63275 2022 (print) | LCC TA418.9.C6 (ebook) |
DDC 620.1/18—dc23
LC record available at https://lccn.loc.gov/2021031672
LC ebook record available at https://lccn.loc.gov/2021031673

ISBN: 978-0-367-75260-6 (hbk)
ISBN: 978-0-367-75261-3 (pbk)
ISBN: 978-1-003-16173-8 (ebk)

DOI: 10.1201/9781003161738

Typeset in Times
by codeMantra

Contents

Foreword

Prior to 1999, I had not heard of sheet molding compound (SMC) nor Probir Guha. From 1999 through 2005, SMC and Probir would have a profound impact on my career and life. Although many people in the heavy equipment and light vehicle industries may not realize it, Probir has also impacted their careers and lives.

This impact statement may sound hyperbolic; I assure it is not.

Probir earned his Bachelor's in chemical engineering from the Indian Institute of Technology, Kharagpur, his Master's in polymer engineering from the University of Detroit, and his MBA from Wayne State University. Despite his extraordinary education, Probir has retained his ability to work with and inspire simple people like me.

In the 44 years since graduating from the IIT, Probir led the development of countless innovations in the field of thermoset composites, including the award of 160+ patents. Many of these patents were awarded to Probir during his 35 years at the ThyssenKrupp Budd Company, where he served as Vice President of Advanced Research and Development. Currently, Probir leads the global development of fiber-reinforced composites for Coats Group PLC.

When I met Probir, and was introduced to SMC, I was leading a new engineering activity at one of the original Big Three auto companies, focused on paint on plastics. The very name of this activity is an indication of how little my organization understood SMC. This substrate technology does have polymers as their backbone, but they are not thermo "plastics." I soon learned from Probir that SMC is a highly engineered formulation of polymer resins, cross-linkers, inorganic fillers, various additives, and of course reinforcing fibers. Like paint, the finished product of SMC is very dependent on the process that converts the formulation into the final molded form (a film in the case of paint).

Most people will recognize the glass fiber-reinforced exterior body components long used on the GM Corvette. In 1999, my company was on the verge of multiple uses of SMC on exterior body components. For many reasons, SMC is an attractive alternative to sheet metal in many automotive exterior body applications; these reasons include complex and deep draw capabilities and reduced tooling costs.

The launch of SMC body components from 1999 through the early 2000s proved to be problematic for the automotive assembly plants. As the Paint on Plastics leader at my company, these problems were mine. The pressure and urgency to correct problems affecting assembly plant performance are intense, relentless, unforgiving, and almost always results in a classic customer-dominant relationship and interaction. It was in this high pressure environment, I learned the most from Probir.

Correcting the performance of SMC required more than just technical knowledge. Fortunately, Probir also provided leadership that extended across his company and mine. Under Probir's leadership, the customer and the supplier formed a unique partnership, allowing real work and real progress. This progress ultimately eliminating the historic root causes of plaguing problems.

Identifying and eliminating fundamental root causes have enabled and widened the use of SMC in automotive body panels. Working with Probir, seeing the power of cooperation and partnership, has made me a better customer and leader.

In 2005, I moved to a new position within my company, but I continued to follow the progress of SMC. Probir and his team did not stop at the elimination of problems with SMC. In more recent times, Probir has led the development of low-density SMC formulations, providing the industry with valuable alternatives to the high cost of lightweight metal parts. In the past 5 years, Probir has been focusing on advancing another form of composites, continuous fiber-based composites. I am certain, with Probir's focus, continuous fiber composites will be another success story in the composites industry.

I really do not know how, why, or if I even deserve it, but during those difficult years between 1999 and 2005, I was able to gain Probir's respect and friendship, both of which are cherished accomplishments and career highlights for me.

As you read this book, I am confident that you will find the technical and the human aspects equally valuable and interesting.

Thank you Probir, for inviting me to participate in this book, I look forward to watching your continued contributions to the successes of the composites industry.

Jim Bielak
President, ACT Test Panels LLC, Hillsdale, Michigan

Editor

Probir Guha's career in innovation began more than 45 years ago after graduating from the Indian Institute of Technology, Kharagpur, and completing his graduate work from the Polymer Institute, University of Detroit in the United States. Immediately before joining Coats, he worked at Continental Structural Plastics, a company that specializes in the design and manufacture of composites materials, where he was Vice President, Advanced R&D. Prior to that, he worked at ThyssenKrupp Budd for 35 years where his work led to many innovations that helped the company establish itself as an international leader in automotive composites.

As Vice President of Global Composites Innovation for Coats North America, Probir currently leads the development of Coats global strategy for fiber-reinforced composites and its center of excellence focused on developing lightweight solutions using textile industry process-driven solutions utilizing continuous fibers for composites.

Prior to that, he served as Vice President Advanced R&D, Continental Structural Plastics, where his accomplishments included building an effective research team that led to the development and launch of the TCA® and TCA UltraLite® SMCs, the development of the first automotive composite-centric Carbon Fiber Recycle system. Additional accomplishments also included coordinating developments between the US and European teams that led to a JEC Innovation Award for the Development of a Lightweight Automotive Hybrid Decklid.

He was also the recipient of the American Composites Manufacturers Association (ACMA) 2019 Pioneer Award, which celebrated his achievements in composite developments for the automotive market. Some of his achievements included being awarded more than 150 patents issued, with many patents pending globally, in addition to being recognized for being a leader in composites innovations over his career.

Between 1998 and 2006, Probir served as the Technology Committee Chairman for the Automotive Composites Alliance (ACA) in Detroit.

Probir is based in the United States and currently resides in the Metro Detroit area with his wife Sri and is the proud father of two sons.

Contributors

Prabir Aditya, PhD
Co-founder and CEO
Machine Learning, Artificial
    Intelligence and IoT
SPRINRIVER Technology Private
    Limited
Kolkata, India

Michael Z. Asuncion, PhD
Senior Materials Scientist, Advanced
    Research and Development
Teijin Group Senior Technical Expert
Continental Structural Plastics, A Teijin
    Group Company
Auburn Hills, Michigan

Cedric Ball
Director, Growth and Innovation
MBE Certified Automotive and
    Composites Industry Consultant
Hexion Inc.
Columbus, Ohio

Jim Bielak
President, ACT Test Panels LLC
Hillsdale, Michigan

Kipp Carlisle, PhD
Engineering and Materials Development
    Manager
Trelleborg Offshore Boston
Randolph, Massachusetts

Mehrdad N. Ghasemi Nejhad, PhD
Professor
Department of Mechanical
    Engineering
University of Hawaii at Manoa (UHM)
Honolulu, Hawaii
Founding Director
Advanced Materials Manufacturing
    Lab. (AMML), Intelligent and
    Composite Materials Lab. (ICML),
    and Hawaii Nanotechnology &
    Renewable Energy Lab (HNL)
Honolulu, Hawaii

Avery Goldstein, J.D., PhD
Founding Partner
The intellectual law firm of Blue
    Filament Law PLLC
Birmingham, Michigan

Vamshi Gudapati, PhD
Manager
Display Innovation Portfolio
Corning, Inc.
Corning, New York

Prateep Guha
Chairman
Machine Learning, Artificial
    Intelligence and IoT
SPRINRIVER Technology Private
    Limited
Kolkata, India

Probir Guha
Vice President
Global Composites Innovation
Coats North America
Charlotte, North Carolina

Kristin N. Hardin, PhD
Doctoral Research Fellow
Department of Materials Science and
    Engineering
University of Alabama at Birmingham
Birmingham, Alabama

David J. Krug III, PhD
Senior Materials Scientist
Advanced Research and Development
Continental Structural Plastics
A Teijin Group Company
Auburn Hills, Michigan

Amnon Levav
Chief Innovation Officer & Co-founder
SIT (Systematic Inventive Thinking)
    A Global Innovation Consultancy
Tel Aviv, Israel

Haibin Ning, PhD
Assistant Professor
Department of Materials Science and
    Engineering
University of Alabama at Birmingham
Birmingham, Alabama

Gajendra Pandey, PhD
Founder and President
SimuSol Design LLC
Atlanta, Georgia

Brian Pillay, PhD
Professor
Department of Materials Science and
    Engineering
University of Alabama at Birmingham
Birmingham, Alabama

Jim Plaunt
AOC (retired)
Technical Sales
Collierville, Tennessee

Dave Reed
Director Emeritus
Society of Plastics Engineers
    (SPE) - Automotive
Detroit, Michigan
General Motors (retired)
Warren, Michigan

Mike Siwajek, PhD
Vice President of R&D
Teijin Group Senior Technical Expert
Continental Structural Plastics, A Teijin
    Group Company
Auburn Hills, Michigan

Introduction

Composites Innovation: Perspectives on Advancing the Industry is a culmination of more than 40 years of experience in the composites industry that all began with a research project at the Polymer Institute at the University of Detroit. Subsequently, the editor has observed changes, challenges, and growth in the industry through the years as a key innovator in the composites world seeking solutions for automotive applications.

In an extremely competitive world, a lot of these changes have been technology-driven. This in turn has attracted top scientific and technical talent to the composites industry.

This is a compilation of the varied experiences and views of the contributors in the field of composites. As a number of the contributors are from the automotive segment, the reader will see a fair share of automotive-related innovations. While we have collected a varied group of contributors to share their thoughts and experiences, no discussion on composites is complete without mentioning Robert Morrison, a pioneer in automotive composites. The company he built, MFG, has a wealth of innovation contribution in this field and is currently led by his son Richard, who continues in the innovative path set by his father in the early 1950s.

As the editor sought to explore new ideas and concepts to further strengthen the capabilities of the industry, he has been able to assemble an accomplished group of global business and technology leaders. Through a variety of views from experienced minds, coupled with new thoughts from well-trained scientists and engineers, this manuscript and the contained discussions provide a panoramic view of our composites world.

The compilation includes a discussion from industry business leaders and the university research community; from the advanced applications in North America and from the emerging markets in Asia; and a timely discussion of advances in the recycling of composites and the emerging smart composites technology. The book provides a landscape of the status of intellectual property and will also encourage the reader to look outside of the conventional world of composites into the use of Artificial Intelligence and a new disciplined approach to ideation and innovation.

The discussion is arranged in five key segments, including how composites fit into our world, the basics of the technology, customer insights, discussions from outside the world of composites, and culminates with a path forward to find competitive and effective solutions in a timely manner.

It is the editor's hope that thoughts and viewpoints and knowledge explored herein will provide an excellent backdrop for current business leaders, researchers, and new entrants to this vibrant community to pioneer new solutions with composites.

Finally, the editor would like to thank all the contributors who enthusiastically came together to share their experiences in compiling this book. It was as if the team was getting together one more time to brainstorm and do the grind to solve an "unsolvable" problem like they have done so many times in their careers, working together. We all agreed that it takes a team and that we all wanted to contribute in our own way to share our experiences to create new solutions and importantly to attract new talent, with new thoughts to take composites in new directions.

The editor would also like to recognize and thank two persons, Allison Shatkin and Gabrielle Vernachio of Taylor & Francis Group, without whose constant help with various details of getting a publication together this work would not have been possible.

Probir Guha

1 Introduction to Composites

Dave Reed

CONTENTS

A composite is a mixture of materials with different properties that combine to provide a material superior to the separate ingredients. This book focuses on composites made from an organic polymer matrix with inorganic reinforcements. This first chapter summarizes the development of common composites as well as their commercial applications and processes. Composites have become the wave of the future because their properties and processes can be varied widely by selecting the type of matrix material, the type of reinforcement, the fiber orientation, and the percent of reinforcement to optimize their properties. Both thermoset and thermoplastic polymers are important in composites. Most composites also include a sizing or coupling agent to assure good adhesion of the polymer with the reinforcement fibers or particles. Sometimes unexpected performance of composites can be achieved with varying combinations of ingredients and processes so intellectual property rights are important and will also be discussed in a following chapter.

Many people in the composites industry might agree that significant advances have come from making good engineering guesses with a lot of trial and error. This empirical approach is not so much from intellectual laziness but from the disproportionate effects that reinforcements and polymer choices and processing variables can have on the properties of composites. A careful balance is needed to optimize improvements in physical properties while minimizing negative side effects. Combining these choices of matrix polymers, reinforcements, sizing, and directional orientation opens a host of possibilities in optimizing properties. Important new developments in Artificial Intelligence and Quantum Computing may accelerate the composites development process moving forward.

DOI: 10.1201/9781003161738-1

A BRIEF HISTORY

Sun-dried bricks made from mud and straw and Mongol recurved bows are successful examples of early composites applications, but the beginnings of modern polymer
composites can be traced to the first thermoset polymers. In 1907, Dr. Leo Baekeland
succeeded in polymerizing formaldehyde and phenol into a hard plastic we know
as phenolic. He named it Bakelite. Dr. Baekeland's phenolic was hard and brittle,
but when combined with wood flour, asbestos fibers, or cotton canvas, the composite became much stronger and tougher. These thermoset phenolic composites were
quickly put to good use in molded electrical insulators and housings for telephones,
radios, and home appliances.

Automobiles were also becoming popular in the early 20th century and used the
new thermoset phenolic composites in several applications that lasted for decades.
Phenolic composites were molded into steering wheels into the 1950s, shift knobs
into the 1960s, distributor housings into the 1970s, camshaft gears into the 1980s,
and ashtrays into the 1990s. Thermoset phenolic composites are still used today in
automobiles for printed circuit boards, brake pistons, brake pads, and clutch friction
plates. Another highly successful composite application in automotive was pneumatic
tires made from bias ply fiber reinforcements in thermoset natural rubber developed in
the late 1800s and early 1900s by Dunlop, Michelin, Firestone, Goodyear, and others.

In the 1940s, composites got another huge boost when it was discovered that other
brittle plastics such as thermoset polyester could be molded by pouring, brushing, and
rolling the catalyzed unsaturated polyester liquid onto fiberglass mat or cloth in an
open mold to become much stiffer, stronger, and tougher as a composite. Unsaturated
refers to polymer chains with multiple double bond sites available for cross-linking.
The catalyst enables the cross-linking between polyester chains to become a rigid
thermoset structure. The applications for these new thermoset polyester composites
came fast and furious in lighter, stiffer surfboards, tougher boat hulls that did not rot
or leak and lighter sports car bodies that did not rust.

Thermoset vinyl ester and epoxy composites were developed because they offered
more toughness than thermoset polyester. Carbon fiber-reinforced epoxy composites offer even greater strength and stiffness than glass fiber (GF) polyester and have
become widely used in sporting goods, fishing poles, golf club shafts, composite compound bows and arrow shafts, and skis and ski poles. High-performance cars are using
carbon fiber epoxy composites for hoods, roofs, doors, deck lids, spoilers, and body
structures for maximum strength, stiffness, and lightest weight. Developments of GFs
and glass bubbles-reinforced vinyl ester composites have continued and can provide
an excellent value in these same applications to save mass, avoid corrosion, and reduce
tooling costs. Now major components of modern aircraft and spacecraft are made with
carbon fiber epoxy composites for wings, winglets, stabilizers, and body structures.

Over the years, further developments in composite materials and processing have
improved surface finish, toughness, and lower weight in automotive body panels. Other
developments have improved the gel coats and resistance to sun exposure and water
delamination blisters in boats. Faster reacting resins have cut molding cycle times by
half and by half again. Some molding processes such as resin transfer molding (RTM)
and vacuum bag molding that once required several hours are now being molded with

sheet molding compound (SMC) in 2 minutes with faster curing resins and processing. These materials and process developments have kept composites on the leading edge of customer expectations for quality, cost, and higher production volumes.

REINFORCEMENTS

As the polymers for composites have progressed so have the reinforcement materials. The major reinforcements being used for composites are GF, calcium carbonate, talc, mica, aramid fiber, and carbon fiber. GF commonly called fiberglass is the most common reinforcement for composites and provides amazing performance for its cost. Fiberglass was first made in 1935 at Owens Corning from silicon dioxide – silica-sand – with added minerals for toughness and strength. The process starts with melting the sand mixture in a furnace and feeding it through multiple spinnerets to make fine hair-like fibers. The fibers are bound together with chemical sizing in various numbers to form thread-like strands as needed for different applications. The strands are then wound onto spools for shipping and handling. Fiberglass strand is used in many forms to reinforce composites – chopped strand, continuous fiber, random mat, thin veil, woven cloth, directional tape. Tiny sand-like glass bubbles and glass flakes are also used for reinforcements to minimize the directional properties of fiberglass. The glass flakes are made by crushing glass bubbles of suitable diameter. One of the most successful composites applications of the tiny glass bubbles has been to mix them in with chopped fiber glass strand in a thermoset toughened polyester to mold automotive body panels almost as light as carbon fiber epoxy composites at much lower cost.

Mineral powders are also important commercial reinforcements in composites – calcium carbonate, talc, mica, etc. These mineral reinforcements offer strength, stiffness, dimensional stability, and cost advantages. Because these minerals are mined, they can have different properties, shapes, and sizes depending on the location of the mines and their processing. Calcium carbonate particles are a powder and have a lower aspect ratio like gravel or sand, so they tend to minimize directional differences in properties of composites – isotropic. For example, calcium carbonate has long been used along with fiberglass in thermoset polyester composites to minimize the directional shrinkage and surface variation of the composite parts caused by the fiberglass orientation. Talc particles have a plate-like shape and have a higher aspect ratio causing more directional properties. Mica particles also have a plate-like shape and have a much higher aspect ratio so tend to add more directional properties to a composite – anisotropic. These inorganic materials do not chemically react with the organic polymer matrix, so the major effects on physical properties come from the average size and shape of the particles and from the chemical sizing or coupling agent that is applied to help the polymer bind to the surface of the fibers or particles.

High-performance organic fibers have also become more important to extend the physical properties of composites. Aromatic polyamide (PA) fibers called aramid fibers were developed by Dupont® out of their nylon PA technology by introducing aromatic monomers during polymerization. Nomex® is a Dupont® aramid fiber that was developed in the early 1960s as a fireproof fabric and quickly became used for fire resistant suits. Kevlar® aramid fiber also from Dupont® was introduced in 1973 as a strong and tough fabric for body armor. Kevlar® aramid fibers have become important in composites applications demanding extra toughness including automotive tire reinforcements and sports gear, high-performance canoes, and kayaks.

Carbon fiber is known for its stiffness and strength and is among the lightest reinforcements for composites. Carbon fibers were first made from carbonized natural fibers in 1860 by Joseph Swan and by Edison in 1879 for the incandescent light bulb. However, it was not until 1964 that strong stiff oriented carbon fibers were first made at Union Carbide in a commercial process by stretching rayon fiber to orient the polymer chains and carbonizing them at 2,800°C. The rayon fiber precursor was soon replaced with polyacrylonitrile (PAN) fiber precursor for even higher modulus and strength developed by Toray® in Japan. This PAN is the most common process to make carbon fibers today. The fine filaments are about one-tenth the thickness of a human hair. They are bound together with sizing to form thread-like strands called tows composed of thousands of fibers in each tow. The strongest and stiffest tows are made with the most fibers per tow diameter and are the most expensive type used for aircraft and aerospace composites. Carbon fiber is becoming more widely used, but it costs more than ten times as much as GF, so it is limited to the most demanding composites applications. The search for lower cost carbon fiber has been the focus of much development for several decades to allow many more applications.

THERMOPLASTICS COMPOSITES

Thus far, we have focused on thermoset composites instead of thermoplastics because thermosets are where modern composites began. Whereas thermoset polymer chains have chemical bonds that cross-link between the chains, the thermoplastic polymer chains are not cross-linked, so the composite can be remelted for easier recycling and remolding. Because thermoplastics do not have cross-linking between the polymer chains, they are generally less brittle and are widely used without reinforcements. But the strength, stiffness, and dimensional stability of most thermoplastics can also be improved for many applications with appropriate reinforcements.

A wide variety of thermoplastic composites have been developed to meet the needs of many applications where enhanced properties are needed. The most common thermoplastics for composites are polypropylene (PP), PA called nylon, and polyethylene terephthalate and polybutylene terephthalate called polyesters. Nylon is a tough thermoplastic that was developed in the 1930s by Dupont® and first used as a fiber in 1940 for lady's nylon stockings. Nylon became important in World War II for parachutes and tents. Most commercial thermoplastics are available in composite grades in increments from 5% to as high as 60% by weight. Over the next few decades, nylon was followed by PP for lower cost and polyesters for reduced stretch.

Adding reinforcements like GFs to thermoplastics greatly increases stiffness, tensile strength, and dimensional stability, while reducing creep, thermal expansion, and moisture growth. These thermoplastic composites are commonly processed as pellets in injection molding much like unreinforced thermoplastics. To make thermoplastic composite molding pellets, common reinforcements like talc, calcium carbonate, mica, or short GFs are mixed into the molten polymer before it is extruded into continuous strand 2–3 mm diameter. Then after cooling in a water bath, the strand is chopped into pellets 3–15 mm long. Composites with longer GFs can provide greater strength to the final molded part but must be processed differently. The GF strands are continuously pulled through the extruder die directly from the spools as the melt is going through so that the GF bundle is encapsulated like making coated wire or the melt can be infused into the fiber bundle. Whether encapsulated or infused, the GFs remain the full length of the whole pellet up to 15 mm long – called "long-glass pellets." To achieve the best physical properties, the long-GF pellets require special care in the injection molding cycle to minimize breaking the GFs during molding.

COMPOSITES MOLDING

Many composites molding processes have been developed over the years for both thermoset and thermoplastic composites. Several of the more common processes will be discussed in approximate chronology of their development because this will help explain how each process improved on previous processes. Thermoset composites can be processed in many ways – open mold layering, chopper glass spraying, vacuum bag molding, matched metal die molding, RTM, SMC, reinforced reaction injection molding (RRIM), filament winding, injection molding, and fiberglass mat thermoplastic molding (GMT) are all used for plastics composites.

Open mold layering is still used in boat hulls and starts with coating the mold surface with a soap-like release film sprayed into the open mold to help the finished part to separate from the mold. This release coat is followed by a precolored catalyzed modified unsaturated polyester gel coat that is sprayed or brushed on top of the release coat to provide a finished smooth outer surface in the desired color. After the gel coat has cured, a first layer of catalyzed, unsaturated polyester is sprayed or brushed onto the gel coal and a thin veil of nonwoven fiberglass is applied into the first coat with rollers to prevent read-through of the following rougher fiberglass reinforcement layers. The following layers can be woven cloth or nonwoven mat. This layering can also be done with a chopper gun that pulls fiberglass strands from spools and simultaneously chops and sprays them into the mold with the polyester resin. This layered composite in the mold cavity is compressed at each layer with a hand roller. Finally, the part is allowed to cure for several hours before removing and trimming.

The vacuum bag molding process is often used to augment open mold layering and is still common for low-volume carbon fiber-reinforced epoxy composites. Until recently, epoxies have required longer cure times and higher temperatures than polyester resin to fully cure. It is important to evacuate all the entrapped air bubbles in the laminate before curing to reduce defects and maximize the composite part properties. A clear plastic bag is placed over the mold with the uncured composite part. The bag is then sealed tight and the air is pumped out, so the composite part is in a vacuum that allows the external air pressure to force out any remaining bubbles as the composite cures. Vacuum bag molding can be further augmented by placing the mold with the uncured composite part in a pressurized autoclave oven to cure faster and force out any remaining air from the composite. "Prepreg" is an additional improvement on this process often used with carbon fiber reinforcements. The prepreg is a thin semi-cured sheet or tape made by impregnating a width of continuous carbon fibers all aligned in the same direction or cross woven either as a sheet or as a tape with liquid epoxy resin that incorporates a "B" stage curing agent. The curing agent partially cures the liquid epoxy into a moldable clay-like consistency that allows the sheet or tape to be handled and placed in the mold as needed. The layers of the prepreg are carefully oriented usually at cross angles to provide directional strength and stiffness where needed and to minimize warpage. When the prepreg layers are applied as a tape, robotics and automation are often used to assure consistent accurate alignment.

Matched metal die molding was developed to improve dimensional accuracy and reduce cycle time. The lower tool cavity half is bolted onto the horizontal bed of a vertically opening hydraulic press and the matching upper core half is bolted to the upper platen of the press. Instead of hand-layering fiberglass cloth or other fibers onto a mold with alternating layers of brushed and rolled-in polymer resin, a preform mat of fibers is placed into an open cavity to cover the whole surface and a measured amount of catalyzed resin mixed with mineral reinforcement is poured over the mat. Then the mating (matched metal die) core is lowered into the cavity that squeezes the resin into the mat and fills the tool cavity. The tool is usually heated to accelerate the cure and reduce the cycle time. The press is closed and the upper half of the mold squeezes the polyester mixture throughout the fiberglass mat. After curing for several minutes, the mold is opened and the rough panel is removed from the tools and edge trimmed and sanded to reduce surface defects.

Resin transfer molding is similar to vacuum bag molding and matched metal die molding with the advantages of both. The reinforcement fibers are preformed into a mat and into a mold cavity, the core is closed into the cavity, and then the thermoset resin is injected. This injection can be assisted with simultaneous vacuum in the mold cavity to evacuate air bubbles and assure the fiber preform is thoroughly impregnated with resin. This process is called vacuum-assisted resin transfer molding.

The SMC process has become the main process used in automotive composites because it is one of the fastest and most consistent processes for thermoset composites. The SMC material is first prepared by mixing unsaturated polyester and thermoplastic polymer modifiers with calcium carbonate ($CaCO_3$) or glass bubbles. The modifiers improve surface finish and toughness. The calcium carbonate or glass bubbles reduce part shrinkage, warpage, and surface waviness caused by the directional orientation of the GFs. The mixture also includes a "B" stage catalyst to thicken the mixture to a clay-like consistency for easy handling prior to the molding operation. The mixture also includes a thermally activated catalyst that cross-links the polyester chains in the heated mold to form a rigid part. The viscous polyester mixture is pumped onto a wide polyethylene film on a long continuous steel wire mesh belt and is then spread evenly with a doctor blade as the mixture and polyethylene film moves on the belt. Then a layer of fiberglass strand is chopped and sprinkled onto the polyester mixture roughly 25 mm thick across the width of the moving sheet. Simultaneously a similar layer of polyester mix is applied to another sheet of polyethylene that is the rolled on top of the first layer and the composite sandwich is compacted between the pair of steel wire mesh belts. The composition of the SMC is roughly 30% GF, 50% $CaCO_3$, and 20% polyester by weight. The thickened composite with the polyethylene film is then rolled up like a carpet and cut to length and allowed to partially cure to an easily handled clay-like consistency. When the SMC roll is ready to be molded, the mold operator unrolls the SMC sheet, cuts the predetermined size pieces, and stacks them on the heated mold cavity in a specific pattern to optimize mold filling. Then as the mold is closed, the SMC squeezes out to fill the mold and cures completely under pressure and heat in a 2- to 3-minute cycle depending on part size and complexity.

RRIM is another successful thermoset composite process that has been used for damage-resistant body panels like painted fenders and door skins. It uses lower pressure and lower cost tooling and equipment compared to thermoplastic injection molding. The RRIM process has been used to mold parts using polyurethane or polyurea polymers for more flexibility and toughness and faster production rates than SMC. The components are kept in separate pressurized heated vessels. One for the isocyanate (MDI) side and a separate pressurized heated vessel for the polyol – polyester or polyamine mix, and a third pressurized vessel for the mixture of milled GFs in a polyol carrier. The milled GFs are made shorter by controlled breaking in a milling process to facilitate mixing and flow through RRIM equipment and tooling. The combined fiberglass and polyol mixture is first metered into a static mixer with the polyol side and then the combined stream enters the main static mixer at the same time as the isocyanate stream enters. The two streams mix by impinging together at high pressure and are forced through the main static mixer that divides and remixes the stream several times as it flows through and into the mold. The static mixer head is bolted directly to the mold because the MDI begins to react with the polyol immediately forming polyurethane or polyurea chemical bonds. The system pressure is set to fill the mold within seconds, and the reaction is complete and ready for part removal from the tool in under a minute.

Filament winding is an important process for molding thermoset composites for golf club shafts, fishing poles, some utility poles, and some automotive leaf springs. The process uses fiberglass strand or carbon fiber yarn from spools, pulls the strand through an epoxy bath, and then winds it onto a mandrel type mold before curing. A more complex process variation can be used to weave dozens of strands simultaneously through a circular ring of rotating spinners to form a reinforced tube similar to the process used to make woven rubber hoses. The parts are typically simple in shape and the fibers are highly oriented for maximum strength and stiffness. The filament winding process has been used in automotive to mold transverse leaf suspension springs for some vehicles including the Chevrolet Corvettes.

Injection molding is the most common process for thermoplastic composites. It uses essentially the same equipment and process as most unreinforced thermoplastics. This common injection molding process is a huge advantage for thermoplastic composites because the equipment and processing knowledge is widespread throughout the plastics molding industry. The injection molding press is more complex, automated, and expensive than the common thermoset composite processes. The tooling is usually made from hardened steel to withstand the extreme pressures of injection molding. The tooling is also more complex and expensive than most thermoset tools because of part complexity and automatic ejector pins and runners. Most injection molding machines have a press that holds the tool cavity and core in a vertical parting line and opens horizontally. The composite pellets are fed into the hopper/dryer by vacuum hoses either from shipping boxes placed near the press or from large silos outside the molding area. Many thermoplastics are hydroscopic, so the dryer is important to remove moisture from the pellets that could cause surface defects or reduce polymer properties in the molded part. The hopper feeds the pellets by gravity into the heated injection molding barrel at the far end from the mold. A large screw is turning inside the barrel to further heat and melt the pellets by mastication. The turning screw moves the molten composite toward the mold. After the mold is automatically closed from ejecting its previous parts, the screw is pushed forward inside the barrel toward the mold by hydraulics or electrics. The flow rate of the composite melt through the sprue at the end of the barrel into the tool and through the tool runners and gates into the mold cavities is controlled by the speed of the screw ramming forward in the barrel. The temperature of the melt is 150°C–250°C depending on the composite plastic, and the tool surface is cooled to maintain 20°C–40°C. After the tool is filled, the barrel screw is forced forward to ramp up the pressure inside the mold to pack the tool as the composite cools and shrinks. The presses are usually rated by clamp tonnage to hold the die halves closed, and 500–5,000 ton presses are common due to the high pressures in the tool cavity. The press platens are opened and closed by hydraulics or electrics. The mold is cooled with temperature-controlled coolant lines drilled through the mold, so the molded part can be cooled for ejection as fast as possible without causing molded in stresses. Finally, the molding press opens the mold and the part is ejected from the tool with automated ejector pins built into the tool. The molded parts can be allowed to fall onto a conveyer belt below the platens or picked off the ejector pins with an automated part removal system or robotics. Then the press closes the mold automatically or when the operator pushes the pair of mold-close buttons. Part-to-part cycle times are commonly 30–90 seconds depending on part size, thickness, and complexity. Some molds incorporate heated runners in the mold so that the parts come out free of the runners and the melt stays inside the heated runners ready for the next shot. Some modern high volume injection molding press rooms contain dozens of presses and operate almost entirely hands free with minimal human oversight. Smaller electric presses molding multicavity small parts in high volumes lend themselves to this level of automation. Injection molding is also used for thermoset composites like phenolics by heating the mold instead of cooling the mold so the thermoset resin cures in the mold. Basically it becomes a highly automated version of the compression molding process commonly used to heat and mold the "B" stage phenolic prepolymer powder.

Glass mat thermoplastic (GMT) molding is another successful thermoplastic composite molding process ideal for larger parts with fewer ribs, bosses or undercuts, and lower production volumes than injection molding. The resulting parts can be much tougher than injection molding because of the continuous GF mat. This process is much less common than injection molding but can accommodate larger parts because the molding pressure is much less. The GMT parts are usually molded with continuous GF mat in PP, but nylon and thermoplastic polyester grades have been made. To make the GMT sheet material, the continuous GFs are dropped onto a continuous belt in a random orientation and the GF mat is needled with gangs of reverse hook needles into an interlocking felt-like mat. For higher strength, the continuous fibers can be oriented for parts like automotive bumper beams. The mat is then fed through a pair of rollers as molten PP is poured onto the rollers. The rollers squeeze the PP into the GF mat and control the thickness of the resulting impregnated mat – typically 3–4 mm. The fiberglass content is commonly 30%–40% by weight. The continuous PP mat is then cooled and cut into sizes as requested by the molder for the specific sizes needed to fill the part mold. The precut GMT is then shipped to the molder where it is placed on a moving belt. The belt takes the precut pieces through an infrared oven where they are melted to a flexible clay-like consistency. As the hot pliable pieces come out of the oven, they are placed into the mold to cover essentially the whole part surface. Generally a vertical press is used to, so the hot piece(s) can be simply laid onto the horizontal open mold. The press closes the mold which squeezes the GMT sheet into the finished shape of the part. The mold is water cooled to quickly cool the part, but cooling must be kept slow enough to prevent molded-in stresses from causing part warpage. The squeezing and cooling are fast especially for larger parts, so fast cycles can be used for relatively large parts. The tooling and presses are much less complex and lower cost than for injection molding. The pressures are much lower than injection molding, so GMT is well suited to mold large parts that require the toughness of continuous GF reinforcement. GMT molding looks like SMC molding and part design limits are similar too. However, GMT is capable of much deeper draws that would require multiple pieces and bonding in SMC. GMT has been used for automotive bumper beams, rear load floors, seat backs and trunks and in cars and SUV's.

THE FUTURE

Composites have continuously improved and applications continue to expand. Certainly, lower cost versions of current composites will continue to expand into everyday applications. However, some cutting edge materials open whole new horizons when combined into composites. One promising new material for composite laminates with incredible properties is graphene. Graphene is made of carbon atoms bound together via carbon–carbon bonds in a hexagonal pattern just one single atom thick. Graphene sheets are also described as single layers of graphite which is a three-dimensional version of graphene. Graphene is the thinnest material known and has the highest tensile strength roughly 100 times the strength of steel. A single sheet of graphene weighs 0.0077 g/m^2. Graphene also has the best electrical conductivity within a sheet and outstanding electrical insulation to adjacent laminated graphene sheets. Graphene composites are being considered for super-capacitors, medical implants, and aerospace.

Carbon and silica aerogels are also breakthrough materials with some amazing properties – sometimes called solid smoke. Both offer high thermal insulation with incredibly low density and huge surface area. Carbon and silica aerogel densities have achieved lower density than air. Aerogels are not only quite friable but also remarkably strong for their mass. Commercial carbon aerogels are now available with an average pore size of 25 nm and a surface area of 700 m^2/g. As materials and processes develop, these aerogels might be combined as the core of a laminated structure using graphene as the structural skin. But the current costs would limit this incredibly light sandwich construction to aerospace applications.

2 Why a Book on Reinventing the Composites Industry?

Probir Guha

CONTENTS

The key elements for composites to reinvent itself and offer today's solutions would be:

- A better understanding of the IP/Innovation Landscape
- How do we best understand the role of each ingredient that makes up a composite – Testing and Characterization?
- Understand strengths & gaps in materials, process, and design technologies
- Evolution of continuous improvement techniques
- Industry collaboration with universities and research institutions
- Implement sustainability
- Incorporate smart textiles and value-addition capability of composites
- Develop a path forward for the composites industry

DOI: 10.1201/9781003161738-2

INNOVATION & INTELLECTUAL PROPERTY
LANDSCAPE FOR COMPOSITES

Between the contributors to this book, there are over 200 patents issued and patents pending globally!

We will spend a lot of time protecting our intellectual properties by creating patents.

Why patents? In a competitive world, it is very important for businesses to maintain their technological edge in the marketplace in a disciplined and controlled basis of intellectual property ownership. For the business entity, a patent or a family of patents allows it to build a technology fence around a know-how that can effectively keep the competition from encroaching in that space. It provides a protected area where the business can thrive without the danger of the key feature that differentiates a material or process or product from being copied by the competition. The study of patents in an area thus allows new entrants to innovate to build a "new mousetrap."

PROGRESS

One rule that inventors have followed over the years is to ensure that feature that they seek to protect by the use of a patent has to be detectable if copied. If an entity can copy the "secret sauce," the patent owner must be able to detect it. When monitoring is not possible, one may seek to protect the IP space through the use of trade secrets.

Chapter 3 is dedicated to the discussion of the technological history and the current patent landscape for the field of composites.

Any story I tell on composites is incomplete without a mention of a person by the name of Joseph Norman Epel.

My love affair with composites started in Dr. Kurt Frisch's office at the Polymer Institute of University of Detroit in January 1974. It had been a couple of months that I was granted a research scholarship by the institute for my graduate studies in polymer engineering. I was overjoyed with the scholarship – fully paid tuition and a stipend enough to cover all of my meager monthly expenses. Then on that fateful day in January 1974, as I was "cooking" an isocyanate terminated prepolymer in the polymer lab, I got the call: "Probir, Doc wants you in his office."

Being summoned by Dr. Frisch had me on edge. What did I do? Was my scholarship in trouble now? As I made my way to Dr. Frisch's office, I realized that others had been "summoned" as well.

Good. I wasn't the only one in *trouble*.

That was the first time I met Dr. Joseph Norman Epel – in Dr. Frisch's office on that fateful day in 1974. Dr. Epel – or Doc – as the world of automotive composites knew him – was to sponsor a research effort at the institute and wanted to meet and select the graduate student who was to work on the program. Well, he selected me and thus began the education of Probir Guha in the conjoining of education and invention and business from one of the best teachers in the world, Doc Epel.

Doc Epel was excellent at getting down to the level of the listener, me, and explaining technical challenges. "Probir, to solve a problem at a molecular level, become the molecule and imagine what it has to go through," he used to say. The first project he had me work on was to address the problem faced with sheet molding compound

(SMC) during a molding cycle – the polymer matrix of the SMC would rapidly drop in viscosity during the molding cycle and lose its capacity to carry the chopped fiber with it during the molding process and the molded part would end up with regions of very low glass fiber content and several other types of molding defects.

My meager knowledge of urethanes kicked in and I was able to participate in devising a path forward for the key SMC formulation. "I became the molecule" and imagined what was required of me and how I could do so.

This early participation led to an application, an innovation, and a first of its kind. The sense of intellectual satisfaction was immense. And it led to a patent – my name was not on it, but it certainly was my first brush with an invention. The invention was disclosed in a patent by Shah, Iseler, and Epel in 1980 [1].

This was an exhilarating experience – creating something new – and made a lasting impression on me and led to many more patents in collaboration with exceptionally talented folks through the years.

Table 2.1 lists some of the patents which have been discussed in Chapters 3, 6, and 15 to illustrate key innovations in composite through the years.

TABLE 2.1
A Few Patents in Automotive Composites

Patent Number	Description	Filing Year	Focus
CA1115299	An early automotive application with continuous fiber	1979	Product
US4707317	An early patent on composite leaf springs	1986	Product
US4867924	Fundamentals of vacuum molding process	1988	Process
US5130071	A process to preheat a molding charge to reduce cycle time	1990	Process
EP0461320	In-mold coating chemistries to minimize surface defects on molded SMC	1990	Materials
US5554252	Heat system for an adhesive bonding process to reduce cycle time	1996	Process
US20030100651	Tough class A SMC	2001	Materials
EP2349667	Induction heating of molds to obtain fast cycle times	2009	Process
US20150376350	Use of treated glass microspheres for low density class "A" surface SMC	2015	Material
EP3389972	A process to recycle carbon and glass fiber from an end product	2016	Process
WO2017153763	Process to manufacture commingled fibers	2017	Materials
US20170354372	Sensors for use in textiles for electronic integration	2017	Materials
WO2020102363	Commingled fiber bundle having integral electrical harness and embedded electronics	2019	Materials
US20200122413	Resin transfer molding with rapid cycle time	2019	Process
US20200224340	Carbon fiber debundling for use in SMC	2020	Process
US20200148279	Multimaterial use for weight reduction on a body panel	2020	Product
US20200139583	Near-net-shape preforms for composite molded parts	2020	Materials
US20200139584	Near-net-shape preforms for over molded composite parts	2020	Process

THE IMPORTANCE OF TESTING &
CHARACTERIZATION OF COMPOSITES

The composites industry has frequently focused on product performance with a consistent focus on how to make the ingredients the "best" possible quality. Often these characteristics were defined by the requirements defined in other noncomposite applications or what the raw material producer had to control in their manufacturing process.

This may have been prevalent but obviously not uniformly true.

Each ingredient in a composite molded part has a role to play in the performance and manufacture of that molded product. For example, an SMC formulation may have up to a dozen different ingredients and a knowledge and ability to qualitatively or quantifiably measure each key characteristic essential. This knowledge not only helps us reduce variability on product and process performance but also provides the means to improve the product.

The list below shows how ingredients of an SMC system affects either steps in the manufacturing process or the final composite product:

- Unsaturated polyester resin
  - Viscosity, fiber wetout, and mechanical properties
  - Chemical kinetics, shelf life
  - Molding cycle time
  - Mold flow and rheology
  - Morphology with low profile additive affecting molded panel dimension and appearance
- Low profile additive
  - Viscosity, fiber wetout, and mechanical properties
  - Morphology with low profile additive affecting molded panel dimension and appearance
- Catalyst
  - Chemical kinetics, shelf life, and molding cycle time
  - Mold flow and rheology, especially during the curing cycle in the mold
- Inhibitor
  - Chemical kinetics, shelf life, and molding cycle time
  - Mold flow and rheology, especially during the curing cycle in the mold
- Filler – calcium carbonate
  - Rheology and flow of the SMC during the molding process
  - Ease of handling of SMC affecting consistency of charge placement in the mold
  - May affect SMC thickening level due to moisture content variability on the filler
- Filler – alumina trihydrate
  - All of the traditional filler attributes
  - Temperature of dehydration will affect efficacy of flame retardancy
- Glass fiber
  - Fiber integrity and bundling – affects flow in mold and mechanical properties
  - Surface chemistry to promote adhesion to resin matrix

- Carbon fiber
  - Fiber integrity and bundling – affects flow in mold an mechanical properties
  - Surface chemistry to promote adhesion to resin matrix
- Zinc stearate or other mold release agent
  - Melt point (if a solid)
  - Migration of release agent to surface of molded part to provide effective release not completely understood
  - Surface energy of molded part with internal mold release agent to understand prepaint cleaning efficacy
- Viscosity reducer
  - Rheology of SMC
- Thickener
  - Thickening mechanism
  - Effect of environment

The important thing to emphasize is that each ingredient in a composite formulation is there to serve a specific purpose. We must try to understand the mechanism of how an ingredient performs its task and select a scientific test method to quantifiably or qualitatively measure how this task is being performed

A further discussion on the subject is included in Chapter 6.

INNOVATIONS TO COUNTER CHALLENGES – MATERIALS, PROCESS, AND DESIGN

To understand the growth of composites from the 1950s and the challenges being faced in recent times, we need to understand the advantages composites had to offer and additional attributes the market has demanded in recent times.

Composites offered the following advantages:

- Weight reduction
- Design flexibility
- Parts consolidation
- Lower tooling costs
- Shorter lead times for product development
- Dimensional stability/integrity
- NVH improvement
- Corrosion resistance

While there were several reasons to design a product in composites, three advantages appear to have assisted the growth of composites in the early years: weight reduction, lower tooling costs, and corrosion resistance. This growth was in the marine and electrical industry and gradually migrated into the specialty, sports vehicle segment. The process of choice in these early days was in open molding hand layup and spray-up and used mostly chopped glass fiber. With higher volume requirements, tightening emissions laws, the molding processes evolved. Several thermoset composite molding processes are shown in Table 2.2.

TABLE 2.2

Composite Molding Processes

Type of Process	Description	Relative Cost			Process Cycle Time	Mechanical Properties
		Molding Equipment	Mold/Die	Raw Material		
Open	Hand lay-up	Low	Low	High	Slow	Medium
	Chopped aminate	Low	Low	High	Slow	Medium
	Compression Molding	High	High	Low	Fast	Medium
Closed	Pultrusion	Medium	Medium	Medium	Fast	High
	RRIM	Medium/High	High	High	Fast	low
	SRIM	Medium/High	High	Medium	Fast	High
	RTM	Medium/High	Medium/High	Medium	Medium/Fast	Medium/High
	Vacuum bag molding	Low	Low	High	Slow	Medium/High

The innovations in composites have been driven by a need to improve cost-effectiveness and derive the best properties out of the composites for target applications and markets. These innovation driving forces have ranged from

- Raw material innovations in chopped and continuous fiber composites
- Manufacturing process innovations to push throughput and first time through capabilities
- Product design targeted specifically to accentuate the advantages of composites
- Incorporate value-adds into the composite that not easily attainable in competing materials
- New, smarter continuous improvement techniques

Three significant innovations are discussed in this chapter. A more comprehensive discussion on innovations in composites are contained in Chapter 6.

For years the quality challenge for SMC automotive body panels were reworking defective parts and rejects due to paint defects. Over a decade, Original Equipment Manufacturer (OEM)-inspired and industry-driven initiatives were adopted:

These efforts were successful to a limit. The improvements were often unsustainable over a long period of time due to process complexity or cost. Often the actions focused on "fixing" the defects once they had been formed.

The visible surface of a painted hood or a door or a decklid or sometimes a Roof is what a sues paying customer sees before purchasing an automobile. A defect that would clearly be visible to the final customer at the point of purchase was not acceptable. This problem needed a solution.

Finally, in the late 1990s to early 2000s, the tough Class A (TCA) SMC was developed and launched. This was an innovation that addressed the core issues that caused paint defects on mold SMC automotive body panels. The crux of the innovation was a novel use of polymer chemistry to design a molecule that was inherently tougher and resistant to microcracks that were the primary cause of paint defects. Due to the promising results seen by the OEM, TCA was launched in several production body panels in a very short period of time – less than 12 months. The quality results as seen in defect counts recorded by the OEM QTA (quality technical assistance) team. The quality improvement, often referred to by the OEM chief engineer, as the "urban landscape" is shown in Figure 2.1. Painted defect rates dropped by over 98% and after months of sustained results the OEM discontinued tracking these defects!

The TCA technology was disclosed in a patent disclosure by Guha, Siwajek, and Yen in 2001 [2].

This application was made in 2001 with the key objective of eliminating paint process defects on SMC body panels.

From the late 1990s to the early 2000s, stamped aluminum and cast magnesium were starting to make an impact in high-volume automotive applications and were gaining acceptance as viable lightweight materials. Aluminum had significant challenges – dimensional inconsistencies caused by post stamping spring back issues and a higher propensity to dents and dings from impact – but it provided a higher than 40% weight reduction compared to stamped steel parts. In the case of magnesium, the weight reduction capability was even slightly higher than aluminum and was being widely accepted even with stringent requirements on how magnesium had to be handled in magnesium casting plants to avoid fires. Glass fiber-based SMC parts fell short of the weight reduction capabilities of aluminum and magnesium.

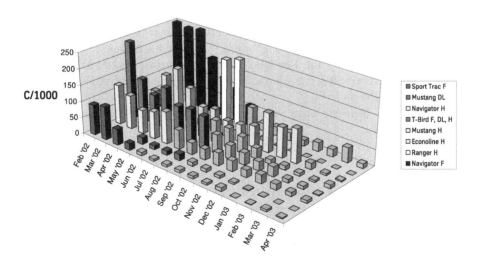

FIGURE 2.1   Defect reduction by using TCA-SMC in automotive body panels. (Courtesy CSP-Teijin.)

ENTER CARBON FIBER

Carbon fiber-based composites afforded much higher weight savings potential than aluminum or magnesium or glass fiber SMC. However, continuous carbon fiber products – unidirectional tapes and prepregs – had a very significant total cost challenge at this stage to break into the mainstream of automotive application. The SMC industry with the support of some key OEMs took to developing carbon fiber-based SMC for automotive applications. Among many challenges to the use of carbon fiber products in high-volume mainstream applications, the two that floated to the top of the sea of challenges were the ability to "wet out" chopped carbon fiber bundles in a conventional SMC machine and total cost requirements still seemed to be out of reach.

Shortly thereafter, a second very key product design-based innovation showed the path to establishing composites as a viable alternative to aluminum and magnesium. The use of hybrid fibers in applications where the molder could use carbon fiber in a cost-effective manner and achieve very significant weight reduction. The approach was still based on chopped fiber as was customary in SMC, but a significant gap had been closed with the promise of additional improvements to come in the future by transitioning to the use of continuous fiber.

Innovations in composite materials, process, and product through the years that have helped maintain the competitiveness versus stamped and cast metals have been discussed in more detail in Chapter 6.

EVOLUTION OF CONTINUOUS IMPROVEMENT TECHNIQUES

Over the years, the Tier-1 automotive suppliers, pushed, cajoled, and led by the automotive OEMs, have used various forms of continuous improvement techniques. The OEMs had a group of experts in their purchasing structure under the banner of "supplier technical assistance" (or similar) to help infuse these new techniques into the realm of Tier-1 expertise to improve their manufacturing systems in a competitive manner. This was a significant effort against the backdrop of innovation efforts that were already in place for the Tier-1s. While the internal innovation effort provided a way for the Tier-1s to differentiate themselves, the OEM-driven continuous improvement drives were to build up the Tier-1 supply base and take the industry to a different level.

A few – and there are many more – of the techniques and methods used are described below.

FACTORIAL DESIGN

A full fractional design will identify all the possible combinations for a given set of factors. In view of the fact that most industrial experiments usually demand a significant number of factors, a full factorial design results in performing a large number of experiments. This effort requires a great deal of discipline and cost. And was often considered to be too complex and time-consuming. However, I must admit that I myself have used factorial designs – and yes, while it could be painfully slow the outcome and conclusions from the effort were extremely reliable.

Taguchi Method

Dr. Genichi Taguchi has developed a new experimental strategy, Taguchi method. This technique was "faster and simpler" compared to the classical factorial design approach. The basics of the method assumed that multiple factor interactions were of less importance in real-life improvements and thus greatly reduced the number of alternatives that needed to be studied or experimented compared to the full or even the fractional factorial design of experiments. This technique was based on an assumption of basing the experimentation on a list of "control factors." A greatly simplified experimental design based on various "statistical assumptions" would be created, and the best part was that due to the simplicity of the design all data could be collected from the manufacturing setup itself. Undoubtedly, there were some production disruption as the team would be tweaking dials and settings to create the boundary conditions required for the study.

This technique came to be used widely. But then, the market wanted faster responses and fewer production disruptions. This led to the teams "skipping steps" so much so that often the team would skip the tried and true ANOVA (analysis of variance) table and just look at level sum outputs and make conclusions.

Increasingly, the customer, the factories, wanted more reliable improvements – FASTER.

Faced with this need, The Budd Company introduced the SMC Consistency Program

The SMC Consistency methodology grew out of the fundamentals of regression analysis and was widely used and reported on in conferences by The Budd Company.

The heart of the technique relied on data and the stepwise regression analysis of a mass of seemingly "unconnected" data to identify the "significant few" to improve process and material for a desired improvement in outcome. This was a perfect fit. Well almost.

A case study of how this technique was successfully used was published and presented at an SPE-ACCE Conference in 2006 [3].

However, new techniques of looking at information and being able to rapidly analyze and understand the information has emerged. The composites industry would do good to make use of these new techniques. If we do not do, so we get left behind! Later in this book, we have two discussions on two such emerging techniques:

- The use of artificial intelligence in continuous improvement. The discussion will include how organizations have adopted artificial intelligence for rapid improvements in methods and processes to leverage organizational benefits across industries.
- A second technique that is starting take hold is based on the "Inside the Box" approach to innovation, which is the basis for both the SIT (systematic inventive thinking) method and the homonymic company. From its inception in 1995, in Tel Aviv, the SIT method has been unique in its structured and disciplined approach to ideation and innovation. SIT stems from a method named TRIZ – originally developed in the former USSR and the acronym stands for "theory of inventive problem solving."

Over the years, we have seen an evolution and progression of continuous improvement techniques. Each individual improvement has improved the capabilities of the composites industry. And now artificial intelligence and SIT could well be the next quantum leap in continuous improvement techniques that takes our ability to RAPIDLY improve to the next level.

UNDERSTAND THE CUSTOMER

My 40+ years in composites and in automotive has revealed that like any other industry segment, automotive is very competitive. The sheer volumes and longevity of programs can be very lucrative in the long run which, in turn, makes the arena extremely competitive. The customer is under relentless pressure from the market to constantly maintain competitiveness – i.e., lower cost; improve quality; and now increasingly, a need for environmental friendliness. The supplier community will require sustained strength in engineering, science, and data analytics skills, but these are the conventional strengths. Suppliers also need to cultivate skills for better communication and a discipline of objectives driven action plans.

Of all the long-term challenges that an industry faced, the battle to meet the customers' quality expectations for a composite body panel was a long drawn out affair. It tested strength and character. For me personally it was a learning experience.

Through these challenges, a company, a team, and a person cannot lose objectivity. In the midst of this multiyear struggle, I would often think of one of Doc Epel's teachings – "understand the objective, do not act till you know the objective"!

Understanding the underlying problem(s) helps define the objective accurately. Sometimes long standing challenges led to confusing the objective. Leading to personal skirmishes. Using statistics to prove a certain point of view without delving into the mechanics, the underlying cause of a weakness. One could hear desperate rants of "our product meets all the quality criteria"; "the quality issues are being caused after we ship our product"; and the classic "they are not paying us enough to meet that requirement."

The important point to keep in mind is that the final customer who buys the finished product from the OEM sets the performance criteria and that evolves. The market creates the pressure on the OEMs by changing their product buying patterns and in turn the OEMs look to the supplier for improvements.

It is not personal. It is market driven. We are all in it together.

My key observations on these nontangible, but very important requirements, are

- Listen to the customer
- Don't sell a product – solve a problem
- The solution to the problem will require multifunctional inputs. It is important to listen to all stakeholders.
- It is not personal
- In automotive you are in for the long haul

These are some of the key elements to winning business in a very competitive world. A section in this book is dedicated to the discussion of fostering a winning relationship with customers, the important space between market and sales – business development, and how cutting edge technology can be a recipe for growth of a company.

A CASE FOR ENDURING INDUSTRY–UNIVERSITY PARTNERSHIPS

The composites industry will need to continue and improve its partnerships with universities and research institutes. Indeed, my own career emanated from a relationship between The Budd Company and the Polymer Institute at the University of Detroit. Over the years, I personally have been associated with some form of involvement in industry–university/research institute collaborative efforts – with Technion of Israel; the Indian Institute of Technology, Kharagpur; Oakridge National Lab (ORNL); University of Western Ontario; the University of Stuttgart; and the University of Alabama in Birmingham later in my career. All of these relationships brought in attributes, speed, and thinking which otherwise may not have been possible.

Earlier in this chapter, I mentioned the collaboration between the University of Detroit and The Budd Company on a material development to improve the flow consistency of the molding compound. Interestingly, the thickening system innovation not only remained in the company's product portfolio for many years but derivatives of this base chemistry led to other innovations ultimately leading to the TCA system, which has also been discussed in this chapter.

Soon after I joined The Budd Company, I found that Doc Epel, who was affiliated with the Technion University in Israel in an advisory capacity, would invite a professor from the university, Dr. Dov Katz, to travel to Detroit and mentor the young cadre of The Budd Company's chemists and engineers advising us on specific development programs we were working on. This was a unique manner in which the company was promoting innovative thinking within its development community by promoting direct association with the best a renowned university had to offer in innovative thinking.

One more example of an industry–university partnership that I was involved with evolved during a visit to my alma mater, IIT, Kharagpur, in the mid-1990s with a senior member of our company's R&D team, Dr. Mike Siwajek (who you will be meeting later in this book), to review some work we were conducting in the Institute's Chemical Engineering Department. During the visit, the professor, Dr. Neogi, very kindly gave us a tour of the various research laboratories in the department which included a trip to the department's Plasma Treatment laboratory. Dr. Neogi engaged us in a discussion of how plasma treatment affected the surface energy of various substrates. The good professor was looking at various substrates for a whole host of reasons and composites were not in his scope of work. Long story short, Mike and I brainstormed some more that evening and decided to kick off a short investigative study on the surface energy of composite substrates and how it would affect the performance of painted composite parts and adhesively bonded composite assemblies.

- What kind of plasma treatment affected the surface of our composite substrates?
- Was the effect significant enough? What was deemed significant?
- Did the change hold through the normal period of the production process and weekend shutdowns?
- How did it affect adhesive bonding and paint adhesion?

So many questions! It led to an excellent cooperative effort between the university and the company separated by ten time zones.

The study showed that plasma treatment of composite substrates led to phenomenal improvement of surface energy of composite substrates. And our standard DRW treatment (Dry Rag Wipe!) prior to the adhesive bonding of molded products was replaced with a completely automated plasma treatment technique yielding a more consistent product. Very quickly this became the norm in our production process for all products. The basic technology was described in a patent by Guha, Siwajek, Haskell, and Sudarsan Neogi in 2012 [4]. The development originated from viewing an unrelated effort at a university followed by work jointly conducted with the university and then implemented successfully by the industry.

In the recent past, the Oakridge National Laboratories (ORNL) collaborated with a major chemical company in an effort to develop a process to develop a low cost carbon fiber. Traditional carbon fiber utilizes polyacrylonitrile as the starting polymeric material to produce carbon fiber. The goal of the joint development program was to use a polyolefin feedstock to manufacture carbon fiber. The new production process was expected to

- reduce carbon fiber production costs by 20%
- reduce the total carbon dioxide emitted in the manufacturing process by 50% by dramatically cutting the amount of energy required.
- increase the yield of the carbon fiber dramatically from 50% to 70% compared to the amount of precursor input into the process

Actual carbon fiber was produced, patents were published by Naskar, Hunt, and Saito in 2015 [5]. But the exact process did not reach commercial fruition in the expected form.

What a novel idea – I fervently hope this approach leads to a lower cost carbon fiber in the future. It would certainly be a giant move forward for composite applications in environmentally friendly lightweight products.

The industry has leveraged the fundamental knowledge base in universities and in research laboratories over the years.

Working closely with the universities will help us alter the interfaces adequately so that we are able to obtain useful solutions faster and the emerging crop of talent joining the fray are ready to teach those entrenched in the industry a few new tricks!

There is a chapter dedicated to a wider discussion on the subject of industry–university–research institution collaboration later in this book.

THE MARKET IS DEMANDING SUSTAINABLE, RECYCLABLE, & CARBON FRIENDLY SOLUTIONS

Environment, sustainability, and recyclability have all gained immense importance over the years. Undoubtedly metals – steel and aluminum – have an ecostructure that meet the sustainability requirement and this fact is well recognized.

Over the years, post-manufacturing thermoplastic composites have been successfully recycled. Not so for thermoset composites. And a product's end-of-life recycling challenges have been daunting.

There have been developments in thermoset composites to incorporate bio-based material and recycled material in formulations. In 1997, Fisher disclosed the use of recycled PET (pop bottles) to manufacture low profile additives for use in composite formulations for the purpose of improved dimensional stability and control [6]. In 2012, Guha, Siwajek, Hiltunen, Pflughoeft, and Swati Neogi disclosed how to use bio-based ingredients in composite formulations [7].

While composites have a very good value story in terms of reduction of carbon emissions, significant recycling and recyclability still elude us despite some of these early efforts at recycle and bio-based ingredients inclusion in composites formulations.

In a world that is increasingly conscious of the need for a "safe" environment, the composites industry needs a solution that is effective both for carbon emissions and for sustainability.

Over the years, thermoplastic composites have been effective in reusing its post-manufacturing waste but have lacked a solution for the end-of-life waste. Thermoset composites, often with edicts from the OEMs, have made sporadic forays into post-manufacturing waste reuse. The most significant of which was an effort to recycle SMC waste as filler back into the SMC stream. But over a period of time, despite pilot plants being commissioned on either side of the Atlantic, this effort could not be financially sustained.

Like any business, any recycle process or system has to be viewed as being able to sustain itself financially. Figure 2.2 shows a process and product flow from

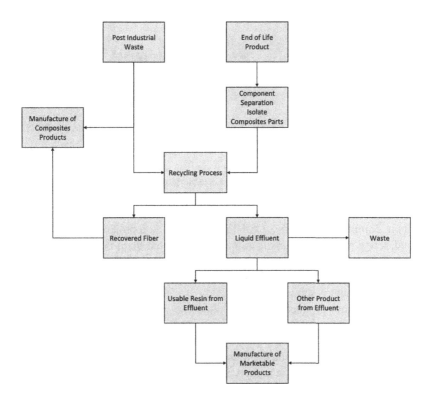

FIGURE 2.2   An environment-friendly recycle process.

FIGURE 2.3   CSP-Teijin pilot carbon fiber recycling 50-liter reactor. (Courtesy CSP-Teijin.)

post-industrial and end-of-life back into a viable product. This entire flow has to be cost-effective.

In the last several years, there have been an increasing number of innovations and patents in composites recycling. CSP-Teijin under took a project to recycle composites and derive product streams from process such that the products could compete with virgin material both in performance and cost. The development was discussed in a patent by Guha, Siwajek, and Krug in 2016 [8]. The pilot reactor set up to prove out the development is shown in Figure 2.3.

The carbon fiber produced from the pilot line was used to manufacture a fully functional automotive decklid displayed earlier in Figure 2.4. The part was awarded the Best Innovation Award at the 2018 JEC Conference in Paris, France.

Efforts like these need to be continued and it is very heartening to note that over the past decade there has been many start-ups that have merged to do exactly that – set up viable recycle businesses. We have decided to dedicate an entire chapter to the subject of Composites Recycling in Chapter 12.

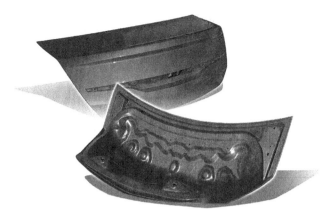

FIGURE 2.4 CSP multi-material Decklid – JEC innovation award winner. (Courtesy CSP-Teijin.)

SMART COMPOSITES

The technology exists today to make the automotive smarter. Figure 2.5 shows a simplified logic diagram of how an automotive can collect information, analyze the data, and allow the vehicle to not only make decisions but also learn from the information as it continues to collect and analyze more information. There exists today data collection sensors that can be incorporated into the molded part. A whole new aspect of composites that will be used in high-volume manufacturing is evolving.

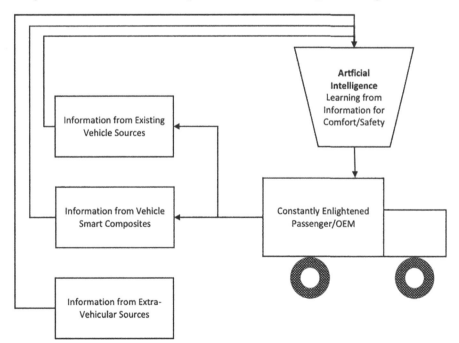

FIGURE 2.5 Smart composite – Artificial Intelligence use in automotive.

This is an emerging field and it is expected that there will be significant innovations in this area in the next few years. Chapters 13 and 17 have been dedicated to the discussion of smart composites and application possibilities and are included later in the book.

REFERENCES

1. Shah, V.C., Iseler, K.A., and Epel, J.N. - Patent CA1079890- thermosetting unsaturated polyester-unsaturated monomer filled composition comprising organic polyisocyanate metallic (Hydr) oxide thickening system.
2. Guha, P.K., Siwajek, M.J., and Yen, R.C. - Patent US20030100651- Reinforced polyester resins having increased toughness and crack resistance.
3. http://www.temp.speautomotive.com/SPEA_CD/SPEA2010/pdf/TS/TS1.pdf.
4. Guha, P.K., Siwajek, M.J., Haskell, B.A., and Neogi, S. - Patent US 20130136929- Plasma treated molding composition for modifying a surface thereof.
5. Naskar, A.K., Hunt, M.A., and Saito, T.- Patent US 20150267322- Method for the preparation of carbon fiber from polyolefin fiber precursor.
6. Fisher - Patent WO 1997028204- Low profile additives for polyester resin systems based on asymmetric glycols and aromatic diacids.
7. Guha, P.K., Siwajek, M.J., Hiltunen, M., Pflughoeft, S., and Neogi, S. - Patent US20140329964- enhanced thermoset resins containing pre-treated natural origin cellulosic fillers.
8. Guha, P.K., Siwajek, M.J., and Krug D.J. - Patent EP3389972- Recycling carbon fiber based materials.

3 Patent Landscape and Inferred Technology Trends

A.G. Goldstein

CONTENTS

INTRODUCTION: BACKGROUND AND MOTIVATION

The properties of composites are detailed from a historical perspective. The modern developments in the field are provided through an analysis of patent activity. Trends in composite development, in general, and the status of the field in particular are inferred from patent classification data used to generate an S curve plot.

A BRIEF HISTORY OF COMPOSITES

Composites are both critical to the next generation of transportation and at the same time ancient. The inclusion of stramineous and particulate plant materials to improve the properties of mud bricks and dried clay objects appear to have been known in disparate parts of the ancient world. These uses are known from archeological sites including those throughout the Levant [1], Africa [2], and China [3]. The use of materials such as straw and sawdust should be viewed as forerunners of the current usage of man-made fibrous and particulate inclusions in modern composites. Still more sophisticated uses of inclusions were known to the ancients include the recognition

DOI: 10.1201/9781003161738-3

that controlled loading of quartz granules improved the elastic modulus of fired clays [4]. As the firing of clay is an irreversible chemical reaction akin to the cure of a thermoset, millennia ago chemical reactions were employed to strengthen composites, even if the mechanisms were not well understood.

Until the 20th century, composite development was limited largely by naturally occurring materials for both matrix and inclusion materials. The combination of synthetic resins developed in the early 1900s coupled with fiberglass led to the first modern compositions that afforded high strength to weight ratios and the ability to produce articles having complex shapes. A dramatic drop in the price of fiberglass was noted between the silk and glass fiber dress displayed at the Columbian Exhibition, Chicago, 1893 by Edward Drummond Libbey [5] and the continuous multistrand fibers developed in 1935 [6]. Aircraft designers quickly recognized the potential value of these materials [7]. An early automotive innovation in lightweighting was the Ford Motor Company "soybean car" made of soybean fiber-reinforced phenolic resin panels on a tubular steel frame [8].

World War II saw the mass production of radomes and aircraft engine nacelles that took advantage of the high strength-to-weight ratio, radar transparency, and corrosion resistance of fiberglass composites [9]. Lightweight composites thereafter expanded for several decades into a variety of fields including sporting goods, pleasure boat hulls, consumer goods, and building materials [10].

While fiberglass came to symbolize technology, it had a limited impact on mass automotive production owing to properties that were still not competitive with steel. The exception being low volume sports cars known for their iconic designs that fiberglass composites made possible. Perhaps the most successful of which was the Chevrolet Corvette series.

The steady improvement in resin properties when combined with new generation of fibers and fillers starting in the 1970s gave rise to modern classes of composites that were finally superior to metal in many structural applications and at the same time provided weight savings. These new inclusion materials included polyaramid fibers, carbon fiber, and microspherical fillers.

Ongoing efforts including providing high gloss surface finish to composites, conductive composites, and joining such components to disparate materials, all while continuing to lightweight components characterize current innovations in automotive composites.

PATENT ANALYSIS AS SOURCE OF TECHNOLOGY TRENDS

Patents represent a bargain struck between a government and an innovator: in exchange for a limited term monopoly granted by the government, the innovator shares with the world through the published patent how to make and use the innovation. Competitors can learn from the patent publication and in an effort to avoid patent infringement and/or payment of royalties to the innovator are incentivized to create additional innovations, and in the process, to propel the technology forward. Patents are especially prevalent in fields such as composites, where a competitor can perform an analysis and determine the constituents. In contrast, trade secret status can be maintained for internal manufacturing processes or precursors that cannot be determined from an analysis of the commercial product [11]. As a result, in the field of composites, the patent reference data analysis can provide insights into the maturity of this technology and offer strategic insights into the future of composites.

PATENT ANALYSIS METHODOLOGY

Patent systems around the world require that a utility innovation be the subject of a utility patent filing either prior to disclosure to the public, or at least within a 12-month grace period. This is in contrast to design patents that are filed to protect the artistic aspects of functional objects. Hereafter, the term "patent" is intended to refer to utility patent references that detail a utilitarian innovation. As a result, patent filings typically predate commercial activity. Whether an individual patent ever protects a commercial product from being copied by a competitor depends on a complex combination of technical and commercial factors. Additionally, patent applications are filed for various reasons that include creating a defensive publication, professional advancement, or marketing that further complicates a narrowly focused evaluation. However, if viewed in a cumulative fashion, patents in a given field can provide information about the maturity of that field.

Figure 3.1 shows a stylized plot of the summation of patents as a function of time for a technology field in which patent filings are routine owing to the lack of suitable alternatives to protect research and development expenditures. To obtain the number of patents allowed in a given time period, one looks to a relevant patent classification. An S curve is commonly observed in a wide variety of technology areas that offers an understanding as to the maturity of the technology. In the "Patent Classification of Composite Materials" section, this methodology is applied to composites and annotated with some historical events detailed above.

Patent Classification as a Source Patent Filing Information

The primary purpose of a patent classification system is to provide a scheme for organizing and ultimately retrieving patent documents by technological subject. By binning patents based on technology covered by the patent claims, related technologies can be located independent of keywords. A classification system is independent

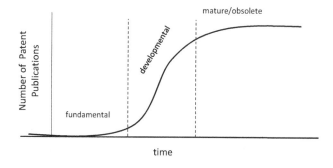

FIGURE 3.1 Stylized Plot Of The Summation Of Patents As A Function Of Time For A Technology Field.

of terminology used by the patent drafter and even accommodates patents written in different languages. As terms to describe innovations are by definition not yet in existence, a patent drafter must often act as their own lexicographer to create new terms to describe the innovation. For example, the telephone was described by early innovators as a "sound telegraph" [12] or "telegraphy" [13] to illustrate the limitations of keyword searching.

There are a variety of coexisting patent classification systems in existence. Many of which are legacies of each individual patent office having to organize paper copies of references. Exceptions to this individualistic approach to classification include the International Patent Classification (IPC) created by international treaty and the Cooperative Patent Classification (CPC) created to unify US and European classification systems with the latter being a refinement of the IPC. Since being implemented in 2013, a variety of additional national patent offices have adopted CPC as their official classification systems. The Derwent® commercial classification system is also in existence.

While a given analysis can be limited to a particular country, region, or common market, it is important to recognize that the assignment of a given patent to a particular class has a subjective aspect that introduces a degree of error into a given analysis. Other sources of error include the subjective aspect of the assigned concordance between older classification systems and CPC, multiple counts to the same technology if filed and granted in multiple jurisdictions, and publications of a patent before grant to create duplicate counts for the same filing. To mitigate the subjective aspects of classification, most patents are provided with secondary or alternative classifications in recognition of the hybrid aspects of most innovations relative to a given classification system. With knowledge of these sources of error and if needed, steps can be taken to cull duplicative results from a search and confirm the desired technology is included in the results.

Curve Segments

The modeling of S curve data is known to be instructive in making business decisions and based on sound analytics [14] yet has met with limited general usage owing to the effort and expense required to generate such curves from commercial sources [15]. The data for generation of such curves should reflect monetary investment, as opposed to hours devoted to innovation (Ibid).

Patent filings represent a form of publication that simplifies the generation of such innovation curves. Within a given field involving professional researchers, the cost of generating a patent filing is quite similar among competitors as the cost factors of research are comparable. These factors include salaries, overhead as to benefits, and needed equipment and supplies. Divergences in cost factors include national prevailing wages, environmental regulations, and equipment cost across the development spectrum from laboratory scale to mass production improvement. As a first approximation, these variations are small over the life cycle of a technology. As a result, the following analysis as a first approximation treats each patent publication as a quantum of investment. One can normalize the quanta of investment per patent publication if desired by applying a weighting as to variable costs if a refined understanding is required. As will be apparent, the unweighted number of patent publications shown in Figure 3.2 provides considerable insights into the field of composites with minimal effort.

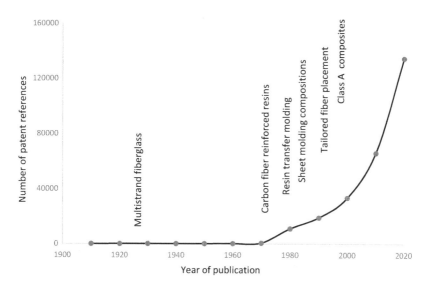

FIGURE 3.2 Partial Sum Plot of Data Series of Table 10.3 With Historical Annotations.

The events represented by a generic S curve (Figure 3.1) based on patent publications include a fundamental phase, followed by a developmental phase, and finally a mature/obsolete phase.

Fundamental Phase

In this earliest phase of Figure 3.1, a small number of patent applications are filed over a comparatively long period of time. The innovations represent a comparatively small investment that often include the disclosure of new classes of materials. University innovations and basic research tend to be more prevalent in this phase than in the following phases. This phase is also characterized by a variety of disparate solutions and parties hedging with innovations on which a technology platform may develop. Patent applicants are allowed to be their own lexicographers in recognition of the fact that standard terminology may not exist for the innovations, and as a result, keyword searching is suspect to locate relevant references. The investment is minimal in this phase and can be generalized as using existing equipment to make some discoveries. As patent terms are finite and only confer at most, 20 years of exclusivity, it is certainly possible that these early innovators have no enforceable patent rights by the time their innovations are a source of serious investment and revenue. Exemplary of this phase would the silk and glass fiber dress displayed by Edward Drummond Libbey in 1893 [5].

Developmental Phase

The developmental phase is centered around a point of inflection in the curve. Others have assigned the labels of growing and mature to regions on either side of this point of inflection [16]. Regardless of the semantics, this phase is characterized by intentional investments of capital that is rewarded with commercial acceptance and revenue. The innovations in this phase bring the saleable product below the customer price point to achieve sales. Innovations that have the net effect of reducing the cost relative to existing mature technologies are common in this phase while often also producing an attribute that existing technologies cannot provide.

As production equipment is installed, a new type of innovation is found in the developmental phase pertaining to the methods of manufacture such as equipment operating conditions and cycles, production facility layout, reduction of scrap, and finding commercial value in waste streams. Over the course of the developmental phase, a single commercial product often differentiates through innovation driven by market demand into a family of products, each tailored for a particular niche of the market.

The patent system is also most notable in this phase. The monopoly conferred by a patent justifies the expense of research and development as the patent holder can maintain high margins on products produces under the aegis of the patent monopoly. As long as there is a high-margin business with some level of sales, there is an incentive to innovate. The product differentiation seen in this phase is attributable not only to customer demand but also to the prospect of extending high-profit margin sales through patent protection to specialty products.

Mature/Obsolete Phase

In this final phase of a product life cycle, fundamental/early development phase patents have expired and the products are now mere commodities. The low margins and competition disfavor investment. Businesses operate with older equipment. Innovations that exist in this phase of focused on highly specialized products and incremental process improvements that nonetheless create cost savings. As these are mature markets, the sales volumes are usually larger than in the developmental phase, even if overall margins are smaller. The high efficiency of the market in this phase and the value it affords customers offers defense against losing sales to new competitive technologies that may offer better properties but does so only at a higher cost needed to recoup research and scale-up costs.

PATENT CLASSIFICATION OF COMPOSITE MATERIALS

The CPC is a system for organizing patent documents, including patent applications and patents, into relatively small sub-collections based on the common technical subject matter. The system uses dots to denote levels of indentation to visually illustrate the nesting of substituent fields into a hierarchy with an increasing number of dots being ever-narrower technical categories. The CPC system has approximately 250,000 categories.

Each classification term has a structure of *Letter1-Number1-Letter2-Number2/ Number3*. *Letter1* is known as a "section symbol" and is a broad field. These are detailed in Table 3.1.

Composite materials are found in section symbol B (Operations and Transport).

Number1 is a two-digit number denoting a "class symbol." **Letter2** denotes a "subclass." The **Number2** is a one- to three--digit number denoting a "group." The "group" is followed by a slash ("/") and **Number3** of at least two digits denoting a "main group" ("00") or "subgroup." A patent office assigns a CPC classification to a patent application or other document at the most detailed level that is applicable

TABLE 3.1

Cooperative Patent Classification Section Symbols

A	Human Necessities
B	Operations and Transport
C	Chemistry and Metallurgy
D	Textiles
E	Fixed Constructions
F	Mechanical Engineering
G	Physics
H	Electricity
Y	Emerging Cross-Sectional Technologies

to the application content. The CPC classifies all technical information in a patent application and not just the subject matter in the published claims. Another benefit of the CPC classification system is that a patent document is classified in all the relevant technical areas, as opposed to being forced into a single or maximal number of classes.

The CPC classification is actively maintained to provide a technologically up-to-date and dynamic classification. As a result, older patents are reclassified to maintain continuity in categories. This is an important aspect in allowing for CPC category data to be used to construct S-curves for composite materials.

Search tools are provided to apply CPC alone or in combination with other information such as keywords, inventor names, applicants, dates, etc. for the efficient retrieval of patent documents containing technical and legal information. The CPC classification is leveraged here for analysis based on the state of the art as to composite materials.

The CPC for composite materials is found in section symbol–class symbol–subclass–group: B32B5. The main group (/00) for this category has the description [17]:

Layered products characterised by the non- homogeneity or physical structure {, i.e. comprising a fibrous, filamentary, particulate or foam layer; Layered products characterised by having a layer differing constitutionally or physically in different parts}

Note

In this group, fibers, filaments, granules, or powder forming or included in a layer may be impregnated, bonded together, or embedded in a substance such as synthetic resin. If the substance of the fibers, or the like, or the impregnating, bonding, or embedding substance, is important, it is classified in the relevant group for the substance

This main group (/00) and the subgroups nested within this hierarchy are detailed in Table 3.2.

A review of Table 3.2 allows one to observe specific technologies that becoming more developed. As a given subgroup, namely a single-dotted (•) category becomes unwieldy as having too many references to retain the ability to efficiently search, double-dotted (••) subgroups are then created to allocate the patents from the single-dotted category. As a general rule, the more levels of categories for technology, one can infer that there is more activity in that category. This general rule can in fact be checked by searching a given category and determining the number of references in that category. By way of example, B32B 5/24 has three, three-dotted subgroups, while B32B 5/20 has no subgroups thereunder.

TABLE 3.2
Cooperative Patent Classification B32B 5/00+ (United States Patent and Trademark Office, 2/2021)

B32B 5/00	Layered products characterized by the nonhomogeneity or physical structure {i.e., comprising a fibrous, filamentary, particulate or foam layer; Layered products characterized by having a layer differing constitutionally or physically in different parts}
B32B 5/02	· Characterized by structural features of a {fibrous or filamentary layer (layer formed of metallic wires B32B 15/02; layer formed of natural mineral fibers B32B 19/02; layer formed of wood fibers B32B 21/02; coated or impregnated fibrous or filamentary layer B32B 2255/02 or B32B 2260/021)}
B32B 5/022	.. {Non-woven fabric}
B32B 5/024	.. {Woven fabric}
B32B 5/026	.. {Knitted fabric}
B32B 5/028	.. {Net structure, e.g., spaced apart filaments bonded at the crossing points}
B32B 5/04	.. Characterized by a layer being specifically extensible by reason of its structure or arrangement {e.g., by reason of the chemical nature of the fibers or filaments}
B32B 5/06	.. Characterized by a fibrous {or filamentary} layer {mechanically connected, e.g., by needling, sewing, stitching, hydroentangling, hook, and loop-type fasteners} to another layer, e.g., of fibers, of paper
B32B 5/08	.. The fibers or filaments of a layer being of different substances {e.g., conjugate fibers, mixture of different fibers}
B32B 5/10	.. Characterized by a fibrous {or filamentary} layer reinforced with filaments
B32B 5/12	.. Characterized by the relative arrangement of fibers or filaments of {different layers, e.g., the fibers or filaments being parallel or perpendicular to each other}
B32B 5/14	· Characterized by a layer differing constitutionally or physically in different parts, e.g., denser near its faces
B32B 5/142	.. {Variation across the area of the layer}
B32B 5/145	.. {Variation across the thickness of the layer}
B32B 5/147	.. {By treatment of the layer}
B32B 5/16	· Characterized by features of a layer formed of particles, e.g., chips, powder {granules (B32B 21/02 takes precedence; layers formed of natural mineral particles B32B 19/00; coated or impregnated particulate layers B32B 2255/04 or B32B 2260/025)}
B32B 5/18	· Characterized by features of a layer {of} foamed material
B32B 5/20	.. Foamed in situ

(Continued)

TABLE 3.2 (*Continued*)

Cooperative Patent Classification B32B 5/00+ (United States Patent and Trademark Office, 2/2021)

B32B 5/22	· Characterized by the presence of two or more layers which {are next to each other and are fibrous, filamentary, formed of particles or foamed (B32B 19/06, B32B 19/048 B32B 19/047, B32B 29/005 - B32B 29/04 take precedence)}
B32B 5/24	.. One layer being a fibrous or filamentary layer
B32B 5/245	... {another layer next to it being a foam layer}
B32B 5/26	... Another layer {next to it also being fibrous or filamentary (relative arrangement of fibers or filaments of different layers B32B 5/12; all layers being fibrous or filamentary B32B 2250/20; two or more impregnated fibrous or filamentary layers B32B 2260/023)}
B32B 5/28	... Impregnated with or embedded in a plastic substance {(not used)}
B32B 5/30	.. One layer {being formed of particles, e.g., chips,} granules, powder
B32B 5/32	.. {At least two layers being foamed and next to each other (all layers being foamed B32B 2250/22)}

COMPOSITE PATENT ANALYSIS

The S curve plot for composite materials can be useful in strategic planning and prediction of market growth. Still greater information is obtained when such curves are constructed for competitive technologies to determine the relative maturity of the technologies. For example, as composite materials begin to compete successfully against metal stampings in the marketplace, resource allocation and incentives shift toward composites with a comparatively slower defunding of metal stampings innovation. An S curve plot is constructed based on the assumptions that within the field of composites that (1) the research investment for each patent is equivalent; (2) a patent publication is an innovation with merit, even if abandoned prior to maturation into a patent; and (3) the number of patents is proportional to the effort being expended in the field.

Selection of a Technology CPC

To examine composites as a whole, usage of CPC group: B32B5 regardless of a subgroup is instructive. The details of this class are found in Table 3.2. This selection of classification at the group level allows for broad trends in inhomogeneous materials, namely matrix materials containing reinforcements to be explored as to technology trends. Another advantage of the selection of CPC value in which hundreds or even thousands of patents exist in a given time frame is that deviations from the underlying assumptions tend to self-cancel thereby reducing the need to expend effort in reviewing the details of the individual patents. While other classification systems are also employed or the filings limited to a finite set of countries or patent applicants, it is instructive to explore global trends to get a sense of the field as a whole.

PATENT DATA GENERATION

The European Patent Office patent search database [18] in an advanced search mode allows a user to select a year or a range of years for patents and publication of patent applications for a given CPC.

The number of references in CPC group B32B5 by decade provides a number of references as detailed in Table 3.3.

The number of references in Table 3.3 shows a steady exponential growth with dramatic acceleration in references seen in the 1970s. The decrease in the number of references in the 1980s is statistically significant. While there was a recession in the early 1980s, this alone cannot explain this decrease in patent references as economic downturns occur at regular intervals. Also, as innovation is subsidized through various forms of tax credits and represents an investment in future earnings, research spending is less susceptible to overall economic downturns. Another possible cause for the reduced number of patent references published during the 1980s is composite production problems. During this era, early composites failed to control for thermal expansion, had poor surface finish, and UV exposure instability. As a result, hesitancy developed in some industries to displace conventional materials with composites. General motors suffered such problems in vehicles produced in the early 1990s [19]. These production composites were based on patent references published in the 1980s. As a result of these setbacks, a reduction in patent publications can be attributed to efforts to explore alternatives for the lightweighting of vehicles relative to composites. While beyond the scope of this chapter, an analysis of patent references during the 1990s, indeed, shows a variety of approaches to address the aforementioned fails as thermal expansion, surface finish, and UV stability.

TABLE 3.3
CPC Group B32B5 Patent References Published by Decade

Decade Ending	Number of References[a] in CPC Group B32B5
1910	63
1920	70
1930	17
1940	38
1950	23
1960	37
1970	389
1980	10,404
1990	7,996
2000	14,436
2010	32,610
2020	68,778

[a] Includes both issued patents and published applications.

Plotting the data series of Table 3.3 as partial sums is shown in Figure 3.2. A steep exponential rise in patent references in B32B5 group is noted from 2000 to 2020 suggesting that composite innovations are proximal to the point of the inflection and therefore in the middle region of the development phase. Figure 3.2 is also annotated with some important composite innovations.

TRENDS INFERRED FROM PATENT ANALYSIS

As Figure 3.2 shows composites to currently be near the point of inflection in the developmental phase, we can expect the next decades to be characterized by enhanced commercial activity. Commercial acceptance largely exists in aerospace and automotive industries as the composite benefits of the high strength-to-weight ratios, dimensional stability, and durability are now commonplace. These increases in investments in composites are a strong indicator that composites are routinely replacing other materials, such as metal stampings. Implicit in this replacement is that composites materials are providing overall value that is below the customer price point so as to displace conventional materials. A review of the number of patents published in subgroups indented within B35B5/00 indicates that trending innovations in composites include specialty compositions with properties such as fire retardancy or conductivity properties added to well-engineered resins, improvements to fiber–resin interfacial properties, selective reinforcement, and complex forms of fiber reinforcements. Equipment and process innovations include carbon fiber separation and improvements in composite cycle time. Composites appear to be following the expected trends for a development phase based on these recent patent references as well as emergent composite recycling technologies. These future trends are addressed in detail elsewhere in this book.

SUMMARY

A procedure is detailed to use published patent references based on the CPC scheme to create an S curve plot for composite materials. Based on the assumptions that (1) the research investment for each patent is equivalent; (2) a patent publication is an innovation with merit, even if abandoned prior to maturation into a patent; and (3) the number of patents is proportional to the effort being expended in the field, an S curve plot is provided for the overall field of composites using CPC group B32B5. The resulting S curve plot confirms that composites are proximal to the point of inflection in the development phase. Based on the generic S curve model, composites are expected to see decades of growth as the field matures.

Recently published patent references show innovations are focusing on niche markets, improved properties, and more efficient manufacturing in support of composites being close to the point of inflection of the S curve model. One can infer that these are areas where investments in research will be productive.

The ease of using patent classification data to generate S curves for composite technology in general has been demonstrated. The approach employed can be extended to specialized composite markets to gain insights into specialty markets. An S curve for a conventional technology being displaced by a composite provides insights as to the rate of competitive transfer of programs to composites. The data used to generate S curve plots are also readily narrowed by geographic region or certain patent applications. Based on the status of composites and the ability of a user to refine plots as desired, insights are provided that can be used in strategic business development and investments.

REFERENCES

1. Homsher, R.S. "Mud bricks and the process of construction in the Middle Bronze Age Southern Levant." *Bulletin of the American Schools of Oriental Research* 368, no. 1 (2012): 1–27.
2. Odeyemi, S.O., M.A. Akinpelu, O.D. Atoyebi, and R.T. Yahaya. "Determination of load carrying capacity of clay bricks reinforced with straw." *International Journal of Sustainable Construction Engineering and Technology* 8, no. 2 (2017): 57–65.
3. Pheng, L.S. "Techniques for environmental control and structural integrity of buildings in ancient China." *Structural Survey* (2004).
4. Kilikoglou, V., G. Vekinis, and Y. Maniatis. "Toughening of ceramic earthenwares by quartz inclusions: an ancient art revisited." *Acta Metallurgica et Materialia* 43, no. 8 (1995): 2959–2965.
5. Sullivan, J. "When Mr. Libbey Went to the Fair." *Bottles and Extras*, Mar.-Apr. (2010): 44–47.
6. Smith, R.A. "History of the use of B2O3 in commercial glass." In *Proceedings of the Second International Conference on Borate Glasses, Crystals & Melts*, pp. 313–322. 1997.
7. Kline, G.M. "Plastics as structural materials for aircraft." *Journal of the Aeronautical Sciences* 5, no. 10 (1938): 391–396.
8. Davis, R. "Henry's plastic car: an interview with Mr. Lowell E. Overly." *V8 Times* (1941): 46–51.
9. Mazzola, M., M. Biagioni, and L. Mazzola. "Composite materials in launch vehicles." *Wiley Encyclopedia of Composites* (2011): 1–9.
10. Nagavally, R.R. "Composite materials-history, types, fabrication techniques, advantages, and applications." *International Journal of Mechanical and Production Engineering* 5, no. 9 (2017): 82–7.
11. Goldstein, A.N., ed. *Patent Laws for Scientists and Engineers*. CRC Press, (2005).
12. Antonio M. United States Patent Caveat No. 3353, (1871).
13. Bell, A.G. United States Patent 174,465 (1876).
14. Nieto, M., F. Lopéz, and F. Cruz. "Performance analysis of technology using the S curve model: The case of digital signal processing (DSP) technologies." *Technovation* 18 no. 6–7 (1998): 439–457.
15. Foster, R.N. *Innovation: The Attacker's Advantage*. Summit Books, New York, (1986).
16. Roussel, P.A. "Technological maturity proves a valid and important concept." *Research Management* Jan.-Feb. (1984): 29–34.
17. Cooperative Patent Classification https://www.cooperativepatentclassification.org/cpc/scheme/B/scheme-B32B.pdf, accessed 2/2021 update.
18. https://worldwide.espacenet.com/advancedSearch.
19. Gardiner, G. Class A composites: A History. *CompositesWorld*. Jan, 19, (2019).

4 Establishing and Maintaining an Effective Composites Development Program

Mike Siwajek

CONTENTS

THE COMPOSITES DEVELOPMENT CHALLENGE

Executing a successful composites research and development program is challenging in any environment, particularly in the automotive market. But as I write this in early 2021, in addition to the usual challenges, there are some daunting obstacles to advancing a technical agenda. The world is in the midst of the Covid-19 pandemic, which will continue to affect the ability to interact face-to-face and conduct business as usual. That aside (and assuming things get under control and back to some semblance of normalcy in the foreseeable future), a shift in the market toward electrification will require significant innovation in a reasonably short period of time. Multiple automakers and even many countries have mandated the elimination of internal combustion engines by as early as 2030.

These hurdles compounded with the usual demands only reinforce the need to have a well-planned approach to building or maintaining a robust R&D program and a well-defined roadmap to achieve success. The base strategy will encompass several elements: physical resources (equipment and capital), building an effective team, global collaboration, and focused leadership. The work plan requires a clear vision of the end goals and a straightforward path to achieve them.

DOI: 10.1201/9781003161738-4

BUILDING THE FOUNDATION

I spent the better part of my formative years intensely involved in athletics, so I am a big proponent of the concept of *team*. I think it all starts there; if a company can build an intelligent, committed, and innovative development team, most obstacles encountered can be overcome. Over my 20+ years in the industry, I have seen different strategies to building a strong composites research group. Early on in my career, we often looked for individuals with experience in our industry, whether they were from competitors, our supply base, or our customer base. There is nothing wrong with experience. But I have also found that sometimes that experience brings the philosophy of, "we've done that or tried that before...it won't work." That kind of thinking can lead to stagnancy and the inability to adapt to a changing market.

We have had tremendous success over the past ten years by targeting scientists from disciplines basic to the composites industry (chemistry, chemical engineering, materials science, and related fields). Hiring directly from universities has actually been very beneficial to our company. Lack of industry experience has not been detrimental. If these people have a strong base in their fields and are well trained in the scientific method, they have the capacity to learn the more intricate details of our particular corner of composites science and apply their skills to our program. Scientists are problem solvers by nature. Members of my team have often encountered challenges similar to those I have faced but attacked it from their own unique perspective and with their skill set to achieve excellent results. I have learned to step back, get out of their way, and only offer encouragement and little bits of guidance when they get stuck. The real trick is to find people with these tools who will be effective members of your team.

There are, of course, some drawbacks to building a team with this approach. First is retention. You spend a lot of time and effort to train new college grads and get them up to speed and fully engaged in your business, but that makes them very attractive to the job market. Paying them well is only part of what it will take to keep them in-house (and that is often out of your hands—that is more on the company and that can often be a significant obstacle). If you can empower these employees and trust them with meaningful projects, they will feel a sense of ownership. Giving them company-wide exposure, letting them sell the idea internally, putting them in front customer, and then allowing them to carry the ball can enhance their growth and job satisfaction as well as benefit the team in the long run. A second downside to having a "green" team is that, especially in the automotive industry, they can often get put in pressure situations where experience would be a huge benefit. The environment can be very intimidating and unforgiving. Putting young, untested individuals in those positions could be uncomfortable for them but, even worse, detrimental to the company. It is imperative they are well trained and to mentor them continually so they can grow in confidence and capability. We have definitely lost a few to bigger and better (or not) opportunities, but at that point, you just take pride in their achievements and start the process over again...and hope that the rest of the team can keep the group moving forward.

Of course, once you have built the team, you must give them the tools to succeed. I was very fortunate in my more recent past to have a CEO who believed that an effective company is a technology-based organization. He felt that if you could create products that the customer wants or needs, that the sales would follow. To facilitate that, we were given a generous budget to build a world-class composites laboratory. All the necessary resources were made available to allow the people to solve current problems, *anticipate* potential problems, and innovate products that made our company an industry leader. Not everyone will have the luxury of a blank checkbook, even for a short period of time. In these instances, it is important to examine the core business and at least acquire the basic tools or access those tools. Developing relationships with your customer base by utilizing joint development and taking advantage of each of the participant's strengths can expand the capability of all teams involved. If the objectives and expected contributions are well defined, this can be an effective alternative to having to build expensive internal infrastructure.

DEFINING THE ROADMAP

You have the team and have given them the toolbox, now you must lay out the plan. In any organization, it is important for the development to be core to the business. This can mean current business—developing improved or more cost-effective products and solving everyday production problems, but it can also mean expanding on current business with a vision for the future. Does the company want to utilize current assets to simply grow the existing business or expand into adjacent or entirely new markets?

It is essential to take inputs from the internal team (Engineering, Sales, Manufacturing, et al.), the current customer base, and the target market(s) to establish the path forward. The goal is to put technology in your company's toolbox to maximize profitability. Research and development should focus on the short-, mid-, and long-term goals of the organization. That is not to say that there is no value in basic research (a.k.a., "Skunkworks"). If you have built an innovative and talented team, a lot of value can be realized by giving those people the freedom to create.

It is also crucial in the development process to understand what it takes to get technology to market. Often something looks very promising in the lab, only to run into issues with the ultimate manufacturability. Going from the lab bench to production scale is not a trivial process. The team should have a solid understanding of the manufacturing process and be familiar with the company business outside of the laboratory. Are the raw materials available at scale and do they fit safety quality control standards? Can the product be made without significant capital investment? Does the end product meet requirements for fitness for use and all customer standards? Is the budget adequate to fund all of the activity necessary? Most important, will the customer ultimately pay for it? Especially in the automotive industry, cost is usually King.

In the composites industry, we have a lot of technologies that many consider "old." I have spent my whole career developing sheet molding compound (SMC) and the entire time hearing that it is falling out of favor and is a dying industry. But we believed that SMC offered value to our customers and we set out to solve the problems that spurred those dire predictions. The development of Tough Class A® SMC (TCA®) was a game changer for us. It immediately eliminated an enormous scrap problem and headache for our plants and our customers' factories. But even more important, it reinvigorated our market. We have since built upon that technology, creating a low-density version (TCA Ultra Lite®) which has proliferated usage and helped us reach a broader market.

There is a lot of "what is old is new again," not just with SMC, but also processes such as resin transfer molding (RTM) and pultrusion—technologies that have existed for a long time but have suffered because of long cycle times and/or limited geometries. Because of innovative thinkers and companies that believed there was value for the market, new chemistries and techniques have been employed to make these materials attractive again. These mature technologies allow companies to use existing capital to sustain, grow, and even possibly reinvent the business.

OVERCOMING ROADBLOCKS

Along this path, there will always be roadblocks. One mantra that my old boss recited time and time again was "Let the data speak." From a pure scientific method standpoint, this is obvious. Define the problem, propose a solution, collect the data, analyze it and you've either developed a solution or you go back, propose a new solution, and repeat the process. It seems like a simple concept, but in a business, opinion and conjecture often cloud judgment. People may have their own agenda or their own problems, and it is easier for them to cloud the situation than to move forward. As scientists, we first must convince ourselves that the data is telling us that we have a solution and then have the conviction to convince others. As someone has spent their entire career formulating, developing, and promoting SMC, I also know that the first thing people blame when there is a problem is the material. One of the first things I teach my team is that it is *always the material's fault*. If they just learn to accept this "fact," it can diffuse a difficult situation. But this is where letting the data speak becomes important. The team needs to be problem solvers—collect the data from development, through raw material manufacturing, through the molding and post-production process. Find the root cause and present the facts (i.e., the data). If the problem truly is the material, we go back and fix it. If it is not, it is our job to help *solve* the problem. It is not to place blame; it is to help eliminate the issue, document the lessons learned, and move onto the next problem.

A method that we have employed to solve issues with SMC formulations as well as manufacturing processes is design of experiments (DOE). DOE is a very effective way to investigate interactions between input variables and determine the optimal solution. SMC can have anywhere from 10 to 25 or so raw material inputs, so looking at all interactions is not realistic. But having a team that understands which inputs to vary and can anticipate the potential outcomes then interpret the data often results in very effective long-term solutions.

Another concern that is almost always encountered is the question of cost. As I stated earlier, cost in the automotive industry is often first on the list. When developing the roadmap, cost must certainly be a consideration. However, if there is merit in the idea, it is imperative to understand the difference between cost and value. Can you ultimately give the customer a product that provides an advantage where a higher cost is justified by the benefits? You must understand the value proposition—if you cannot convince yourself that the product has merit, it is best to move on to another project.

One last roadblock I will mention is the competition. You must first understand who your competition is. We struggled with this in the SMC industry for quite a while. Over the past 20 years, we have seen significant consolidation of our industry.

Several of the bigger players have bought each other out, moved on to focus on other industries, or simply closed shop. What we realized is that we needed to redefine our competition. Fighting over pieces of existing business was not conducive to growth. When we looked at our value proposition and some of the new products we were bringing to the market, we refocused on other materials as our "new" competition. Steel and aluminum have always dominated the automotive market, but composite materials offer many benefits over metals like weight reduction, parts consolidation, ability to mold into complex shapes, tailorable properties and performance, and lower cost tooling. These qualities should allow us to compete for a bigger piece of the automotive pie. I doubt we will ever be as prolific as those materials, but the advances in composites (both thermoset and thermoplastic) make them very attractive to our customers. Also, by redefining the competition, it allowed us to deemphasize another roadblock—"Defensive R&D." Defensive R&D for me is the need to create "me too" products simply because our historical competition developed something new. We focus on our business and our customer base and only develop if there is merit in creating a product. Now we cannot be naïve. As we continue to grow and have success, competition in our own industry can be motivated to follow a similar path and become more formidable. We must continue to innovate so that we remain an industry leader.

UTILIZING PARTNERSHIPS

One way to reduce burden, both on manpower and required funding, is to engage in partnerships. At the simplest level, direct collaboration with supplier partners where bringing innovation to market is mutually beneficial is ideal. Identifying capable suppliers and building a shared trust can allow for fast-tracked development that is not significantly burdensome to either party. I have had some extremely positive experiences with long-term partners where we were able to openly share knowledge, data, and even laboratories resulting in breakthroughs that would not have been possible by either of us on our own.

A current trend has been to engage in partnerships with the automotive OEMs (original equipment manufacturers). Historically, that industry was reluctant to develop in conjunction with the supply base. They were more inclined to ask for solutions to problems, test them out on their own, and then either implement or send us back to the drawing board. But more recently, the OEMs have been open to working on projects together with a few interested and capable parties. It is a true collaborative partnership from the design of the project and creation of objectives all the way to potentially building tools, often with shared funding and manpower. Sometimes there is not necessarily a purchase order for business at the end of the project, but the knowledge gained, and the trust built between the teams is invaluable and has sometimes led to business on latter projects.

Consortiums are another common source for collaborative development. Scientific societies and associations are prevalent in the composites world and have a long history of supporting projects that are considered beneficial to the advancement of the industry as a whole and hopefully also to the individual participants. Many of these require paid membership, potentially contributed funding, as well as in-kind investment of time and/or financing. If the projects are properly identified and planned, these partnerships can be extremely beneficial. They can bring together entities with unique capability and talents and produce excellent results. The drawback is that it is often research that is for public consumption (especially if government funding is involved) and it is possible that you might be teaching your competition. But the benefits of producing meaningful research and developing relationships with partners that may not be direct associates can be worthwhile. I also think there is merit in being a steward for the composites industry.

Academic institutions can also be useful partners for development. Universities often have very specific strengths and capabilities that may not be readily available in a corporate lab. The ability to focus on one particular aspect of a large project can contribute to solutions that might be difficult to identify in an environment that must also support day-to-day functions of a company. Colleges also employ students who are very eager to work on real-world projects and make connections with companies that may employ them one day.

LEADERSHIP

As I mentioned earlier, I was molded by my involvement in athletics. As such, I consider myself a team captain rather than a boss. My job is to define the objectives, help create the roadmap, provide insulation from outside distractions, and let the team get the job done. A crucial part of that function is to help prioritize projects. This cannot occur in a vacuum. The leader needs to be in tune with the objectives of the organization and assure that the projects stay on target. It falls upon the leader to communicate the vision and provide motivation to the team.

WHAT'S NEXT?

In planning for the next several years, I see five major trends in the automotive composites industry: globalization, electrification, lightweighting, mixed materials, and environmental impact.

Globalization is imperative because our customers demand it. The automotive OEMs are moving toward global platforms and, therefore, require common parts to be available in all regions. Globalization presents significant challenges, the first of which is the basic infrastructure to provide just-in-time parts to plants all over the world. There is a lot of investment and complexity to achieve that ability. As a composites manufacturer, that includes identifying and developing local supply for chemicals and other raw materials. With differing regional hazardous materials regulations, the inability for some raw material suppliers to operate globally, and the need for consistency across the entire globe (both in incoming raw materials and our own manufactured products), this is a monumental, but necessary, task. Also, from a development standpoint, it can be extremely difficult and expensive to protect intellectual property. Creating a patent position in multiple countries gets very expensive very quickly. And monitoring IP protection of trade secrets at a local level is logistically challenging. This is definitely concerning, but not a barrier to growth. It is just a topic that needs to be seriously considered.

I have already touched a bit upon electrification. For the composites industry, electrification presents new and prolific markets. The difficulty is that the automotive industry is quickly moving to electric vehicles while still defining the requirements for performance. To us, that means that the target is always moving. It has not been uncommon in the very recent past to develop on the fly at unprecedented speed only to find out the requirements have changed significantly. This often occurs after the material approval process has commenced, so it is very difficult to make major modifications without having to start the whole process over again. Again, electrification presents a very exciting opportunity for composite materials, but the shift will not be without significant challenges.

Lightweighting has been the single biggest growth area for our products. The Obama administration set the 2012 EPA mandate of 54.5 mpg by 2025 which resulted in a huge effort in the automotive industry to reduce weight. While these targets were eventually reduced or delayed, the push toward electrification has further highlighted the need for lightweight materials. This should continue to be an advantage for composites development.

The proliferation of lightweight materials has also led to the effort to combine dissimilar materials to optimize weight reduction. Bonding and joining thermosets, thermoplastics, lightweight metals (e.g., magnesium, aluminum, and high-strength steel) present difficulties due to compatibility with adhesives and coatings, differences in coefficient of linear thermal expansion (CLTE), and problems with galvanic corrosion. While these are problems, they are also opportunities for market growth with effective research and development.

The other, and somewhat more nebulous, topic is environmental impact. As is evidenced by the global shift to electric vehicles and the elimination vehicles powered by internal combustion engines, the environment is becoming a driver for automotive reform. While this is an opportunity to grow the composites market, there are hurdles as well. From an automotive composites manufacture's point of view, volatile organic content (VOC) reduction is the most immediate task. Global standards for lower VOC content are becoming increasingly stringent and present a tall task for traditional thermoset composite materials. The tasks are not insurmountable, but mitigation can mean quantum shifts in formulation chemistry and/or modifications to the manufacturing of composite parts that can adversely affect complexity and cost. Recyclability is becoming a requirement. While it is possible to recycle most composite materials (including thermosets), the infrastructure to effectively accomplish it is nearly nonexistent—especially for end-of-life vehicles. Another term commonly heard is "LCA" (life cycle analysis). LCA is the calculation of the emissions created in the entire life cycle of a product—from the production of raw material all the way through the final part. A few years ago, it was a marketing term but now customers are inquiring about LCA as these companies try to comply with global targets for carbon-neutral status. Composites manufacturers need to quickly react to comply with these new customer requirements in an area that is largely unfamiliar to most. While the reduction of environmental impact seems like a daunting task, it is ultimately a necessary component of social stewardship and is simply the right thing to do. "Climate change" or "global warming" or whatever terminology is used is a scientific reality and, as scientists, it is our duty to do what we can to reduce our contribution to the problem.

SOME FINAL THOUGHTS

Composite Materials Development has evolved over my 20+ year career. The challenges that lie ahead in the next several years will shift our focus and no doubt give rise to new and exciting technology. The value proposition for composite materials will remain attractive to the automotive, heavy truck, marine, aerospace, and consumer goods markets. To conduct effective research and development, it is necessary to build a strong team, give them the resources to succeed, the understanding of the team's objectives, and the path to get there.

5 Testing and Characterization for Composites

Michael Z. Asuncion and David J. Krug III

CONTENTS

INTRODUCTION

The tools presently available to engineers and scientists to characterize, evaluate, measure, and otherwise navigate the wide field of composites are nearly as diverse and numerous as the types of fiber-reinforced polymer composites themselves. Therefore, the authors humbly put forth this chapter to the composite manufacturing community with the intent to concisely describe some of the instruments and techniques they have found useful in their somewhat brief (30 years combined), but certainly informative experiences. The content of this chapter is by no means an exhaustive review of composite characterization but rather intended to highlight key topics that will hopefully aid the reader in their work or to inspire further, in-depth study on particular topics elsewhere.

DOI: 10.1201/9781003161738-5

Materials characterization is an essential element in the formulation, development, and manufacture of fiber-reinforced polymer composites. Identification and understanding of physical, chemical, and mechanical properties allow scientists and engineers to build structure–processing–property relationships that guide the design and quality control of composite materials. The term material characterization encompasses a vast space within the field of material science; therefore, this chapter focuses on the key methods and instrumentation commonly used on fiber-reinforced polymers employed in the automotive and related industries. A keen understanding of material properties and how they change throughout the composite manufacturing process affords the ability to optimize critical properties of a finished part while maintaining a necessary level of reproducibility.

THERMAL ANALYSIS

DIFFERENTIAL SCANNING CALORIMETRY

Common Uses in Composite Materials Characterization: identification of glass transition temperature, melting point, degree of crystallization, extent of cure, oxidation induction time, and material purity

Differential scanning calorimetry (DSC) is the most pervasive thermal analysis technique employed in the polymer and polymer composite industries due to its ease of use and the important characteristic properties identified. By measuring the change in temperature of a sample and reference as a function of temperature and time, the heat flow associated with various thermal events or transitions is derived. Some of these thermal events include the glass transition, melting, boiling, crystallization, and polymerization. DSC provides other useful information such as heat of reaction, reaction kinetics, purity, heat capacity, and oxidative and thermal stabilities.

DSC samples can be either liquid or solid (film, fiber, powder, etc.) and the amount needed is relatively small, 5–20 mg. Samples must be accurately weighed and placed in small pans (typically aluminum, but other metals or ceramics are available as needed) onto which a lid can be crimped or hermetically sealed (to prevent loss of volatiles) if desired. Sample preparation is particularly important for composite materials as they are heterogeneous by nature. Best practices include taking multiple samples and from various areas of a test coupon or real-world part to ensure repeatable and representative data. This ideology extends beyond just DSC and all thermal analysis and should be considered for all composite property analyses including spectroscopy, chromatography, microscopy, etc.

The sealed pan with the sample in it is placed inside the DSC cell on a stage that contains a thermocouple that will measure the sample temperature during the experiment. A reference, typically an empty pan and lid, is placed on the second stage in the DSC cell so that when subjected to the same experimental conditions (time, temperature, etc.) the temperature difference between the two can be measured. A deviation in temperature between the reference and the sample results from the change in either heat capacity or enthalpy associated with a thermal event in the sample. Changes in enthalpy can be endothermic or exothermic depending on the type of thermal transition. DSC data is typically represented by a thermogram where heat flow is plotted against temperature or time. When a crystalline or semi-crystalline thermoplastic undergoes melting the polymer absorbs energy (heat flows into the sample), the DSC measures the temperature difference, and the event appears as an endothermic peak in the thermogram. Crystallization and polymerization are exothermic events and appear as exothermic peaks since heat is flowing out of the sample. Melting and crystallization temperatures are the key properties for processing thermoplastic by injection molding, extrusion, pultrusion, thermoforming, or other techniques. Understanding polymerization temperatures and rates are critical to optimizing the cure conditions of thermoset-based composites.

In most software nowadays, exothermic events are represented by positive peaks and endothermic events are negative peaks, but one should be careful to check the orientation of a thermogram before analysis. The area under these peaks is integrated to determine the heat of fusion in the case of melting or the heat of reaction in the case of polymerization. The latter is useful for determining the degree of cure of a thermoset. The residual heat detected by the DSC of a partially cured thermoset can be divided by the total heat of reaction to represent the degree of cure by a percentage. However, if the sample is a highly filled/reinforced heterogonous composite, the variation in the total heat of reaction of individual samples can be significant, but it may also be possible to pair the DSC data with thermogravimetric data (more below) to compensate for the variation in polymer to the inorganic ratio in the composite samples.

Experimental procedures (time, temperature, ramp rate, atmosphere, etc.) can vary greatly, but most test methods are variations of either ramp or isothermal experiments. A common ramp experiment involves heating at a constant ramp rate (often 10°C or 20°C/min) in an inert nitrogen atmosphere to a temperature above pertinent thermal events such as melting for a thermoplastic or cure temperature for a thermoset, cooling to ambient or sub-ambient (with the aid of a chiller, greatly widens breadth of polymers that can be characterized) at a constant ramp rate and heating a second time through the thermal events. This type of method is called a heat-cool-heat experiment. The first heat will erase the thermal history of the polymer, such as residual stresses from processing, or ensure full cure in the case of thermosets. The second heat provides the inherent properties of the polymers such as the glass transition temperature, T_g; melting point, T_m; etc. The T_g is an important property of polymers and appears as a shift in the baseline due to a change in heat capacity. Below the T_g, a polymer will be glassy and rigid, and above the T_g, the polymer will be rubbery and flexible. This is important when selecting polymers for composite applications with defined operating conditions, e.g., flexible at room temperature or rigid under high heat conditions. The T_g is related to many key properties of polymers including the degree of cure, cross-link density, hardness, crystallinity, and dimensional stability.

The thermogram in Figure 5.1 shows the first and second heats of an extruded polyethylene terephthalate (PET) pellet from a heat-cool-heat experiment ran in nitrogen. Note that the exothermic peak direction is called out in the lower left-hand corner (Exo Up) and that the curves are offset on the y-axis for ease of viewing. In the first heat (solid line; 10°C/min), three thermal events are observed. First, a small endothermic peak appears near 75°C. This is actually enthalpic relaxation (not a melting event) where stresses from processing (i.e., extrusion) or thermal history (long-term storage below T_g) are released as the polymer transitions from a rigid glassy state to a

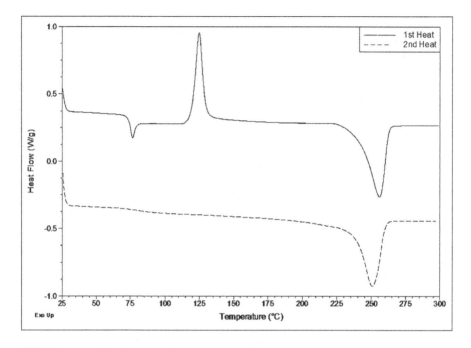

FIGURE 5.1 DSC of an extruded PET pellet showing the glass transition, cold crystallization, and melting; solid = first heat, dashed line = second heat, 10°C/min heating, 5°C/min cooling, in nitrogen.

flexible rubbery state. This is confirmed by the absence of an endothermic peak in the same region of the second heat (dashed line; 10°C/min) and now the subtle T_g can be seen starting near 70°C and ending near 90°C. Since the glass transition occurs over a temperature range, the T_g is often assigned to either the onset or the inflection point of the transition, which in this case is about 80°C. However, it is important to consider the full range of the glass transition when characterizing polymer matrices. The second thermal event in the first heat appears as an exothermic peak with a maxima near 125°C and is associated with cold crystallization whereupon heating of a solid amorphous polymer the chains orient into a more ordered crystalline morphology. After the first heat, the sample was cooled at a slow rate (5°C/min) that allows time for crystals to form and thus no cold crystallization is observed in the second heat. The third thermal event in the first heat is an endothermic peak with an onset and a minima around 240°C and 255°C, respectively, and is related to melting. The thermal history (enthalpic relaxation and cold crystallization) is erased during the first heat and cooling steps, and as a result, only the T_g and T_m are observed in the second heat.

An inert gas (nitrogen, helium, argon) is usually used as the purge gas in DSC experiments to prevent oxidation. However, air or oxygen can be used for oxidation induction time (OIT) experiments where a composite sample is heated in an open pan to a certain temperature (such as processing or operating temperature) and held at that temperature (i.e., an isothermal experiment) until the onset of oxidation is observed as an exothermic peak.

THERMOGRAVIMETRIC ANALYSIS

Common Uses in Composite Materials Characterization: thermal and oxidative stability, filler/fiber content, and degradation behavior

Second, only to DSC in prevalence, thermogravimetric analysis (TGA) is a widely used thermal analysis technique in the field of fiber-reinforced polymer composites. In short, the TGA is a precision balance inside of a high-temperature furnace. It measures the change of a sample's mass as a function of temperature and time in a controlled atmosphere. From these thermograms, one can learn much about the composite's thermal stability, oxidation stability, composition, filler/fiber content, sorption, and desorption.

TGA samples are small (up to 30 mg, but often 10 mg) and can be fully formed composites or any of the starting materials such as polymers, resins, fillers, or fibers. Samples are loaded into alumina crucibles (sapphire, platinum, etc., also available) with or without an alumina lid for volatile or lightweight samples and placed on the thermobalance inside of the TGA, which could be on a hang down pan, vertically supported stage, or cantilever arm. An empty crucible is typically loaded into the corresponding reference position. A purge gas flow is used to control the environment like with DSC and most often inert gases (nitrogen, helium, or argon), air, or oxygen is used. However, humidity-controlled, high pressure, corrosive gas, or dilute hydrogen gas experiments can be conducted under careful guard with proper equipment.

Like with DSC, TGA experiments can be static (isothermal) or dynamic (temperature ramp) or a combination thereof and similar ramp rates are typically used for composite samples (10 or 20°C/min). TGA instruments can reach a maximum temperature of up to 1,600°C, but ≤1,000°C is usually sufficient for working with polymer composites. During a dynamic TGA experiment, the change in mass is plotted against the temperature in a thermogram. Experimental conditions under which mass gains would be observed are less common in the field of composites but could be from oxidation in an oxidizing purge gas or from absorption in a humid environment for example.

In temperature ramp experiments, conducted in an inert atmosphere, mass losses are observed and are caused by different thermal events. Desorption or evaporation of volatiles, such as moisture, solvents, plasticizers, or other low-molecular-weight compounds, is observed at relatively low temperature (e.g., <200°C–250°C). At higher temperatures, the polymer matrix of a composite will undergo thermal degradation caused by depolymerization, random chain scission, and/or side group elimination resulting in a mass loss. Polymer blends or laminates can have more than one mass loss if the constituents have different thermal stabilities. A common measure of thermal stability for polymers and composites is the temperature at which a certain mass loss percentage is achieved; for example, the $T_{d5\%}$ is the temperature at which the sample has lost 5% of its starting mass.

Figure 5.2 shows the thermogram of a glass fiber-reinforced epoxy composite sample ran at 10°C/min to 1,000°C with a purge gas switch from nitrogen to air at 800°C. The weight loss of the sample is plotted against the temperature ramp and shows the main mass loss due to the thermal degradation of the thermoset epoxy matrix with an onset near 340°C. The $T_{d5\%}$ of the sample is approximately 360°C and after the main mass loss the sample weight has dropped 30%. There is a second

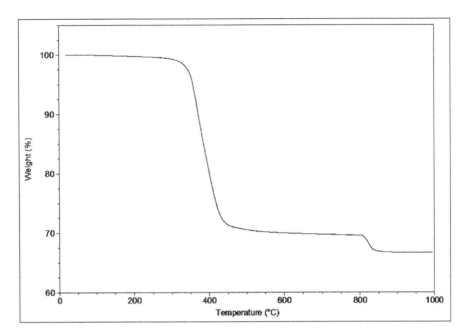

FIGURE 5.2 TGA of a glass-reinforced epoxy composite sample ran at 10°C/min to 1,000°C with a purge gas switch from nitrogen to air at 800°C; $T_{d5\%} \approx 360°C$, fiberglass content $\approx 67\%$.

mass loss at 800°C when the purge gas flow is switched from an inert atmosphere to one containing oxygen, which allows the char formed during pyrolysis to burn resulting in an additional 3% mass loss. The final mass of the sample can be used to derive the fiber glass content since no other inorganic fillers were incorporated into this particular composite formulation.

Filler particles are incorporated in polymers and composites to increase stiffness, abrasion resistance, and thermal stability, or to control shrinkage, rheology, density, and conductivity, or [quite often] to reduce cost. These fillers and additives can also undergo thermal decompositions at elevated temperatures. For example, alumina trihydrate (ATH) is widely used as a flame retardant in polymers and composites because it undergoes an endothermic reaction around 200°C, which both absorbs heat and releases three water molecules for every molecule of ATH thus extinguishing the flame and suppressing smoke.

An interesting extension worthy of note is the combination of TGA with evolved gas analysis or TGA–EGA. By connecting the outgas port of the TGA with a Fourier transform infrared spectrometer (TGA–FTIR), a mass spectrometer (TGA–MS), or a gas chromatography–mass spectrometer (TGA–GC/MS), the composition of the gases released or produced upon heating and decomposition of the sample can be analyzed.

DYNAMIC MECHANICAL ANALYSIS

Common Uses in Composite Materials Characterization: measurement of storage modulus, loss modulus, damping, and glass transition temperature

Polymers are viscoelastic materials because they exhibit both viscous and elastic behavior under deformation conditions. The mechanical properties of a polymer, and therefore fiber-reinforced polymer composites, depend on the rate and duration of applied stresses or strains. At low temperatures and high strain rates, polymers behave like elastic solids, and at high temperatures and low strain rates, they behave like viscous liquids (flow). Dynamic mechanical analysis (DMA) is a technique that applies an oscillating sinusoidal stress or strain to a sample and the resulting out of phase sinusoidal strain or stress is measured. The applied and resultant stress/strain are used to calculate the complex modulus (E*), which in combination with the measured phase difference (δ) can be used to derive the storage modulus (E′) and loss modulus (E″). The storage modulus is a measure of the stored energy and represents the elastic component of the polymer, which relates to the material's stiffness. The loss modulus is a measure of the energy dissipated by molecular motion and represents the viscous component of the polymer. The ratio of E″/E′ describes the damping of the oscillating forces and is equal to the tangent of the phase difference, tan δ.

For the application of fiber-reinforced polymer composites, bars are typically cut from flat molded plaques. The uniformity of width and thickness (more so) are important for the geometric variables in subsequent property calculations. Specimen length is less important because composite samples are loaded into single/dual cantilever clamps or in a three-point bending fixture. When samples are cut from larger molded articles, the rough surfaces must be sanded (and ideally polished) smooth to eliminate any surface defects that could affect the measurement.

A typical test procedure, such as that used to collect the thermogram in Figure 5.3, would be a ramp of 3°C/min from ambient to about 50°C above the sample's T_g at a

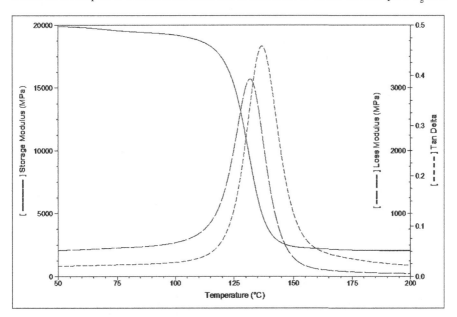

FIGURE 5.3 DMA (3°C/min, 1 Hz) of a carbon fiber-reinforced epoxy composite; E′ = solid line, E″ = long dashed line, tan δ = short dashed line, $T_g \approx 135°C$ (tan δ peak).

frequency of 1 Hz and an amplitude of 15 μm. Figure 5.3 shows the storage modulus, loss modulus, and tan δ of a carbon fiber-reinforced epoxy composite DMA bar loaded in a dual cantilever fixture. The entire glass transition can be assessed by looking at all three curves. The storage modulus drops significantly (an order of magnitude) from 20 GPa to 2 GPa through the glass transition and the onset ($\approx$ 120°C) indicates when the mechanical properties and stiffness begin to decrease. The peak of the loss modulus ($\approx$ 130°C) indicates the highest rate of increasing polymer chain mobility. Lastly, the peak of the tan δ ($\approx$ 135°C) indicates the ratio of E″/E′ and damping is at a maximum and is most often used as the single value of T_g in the literature.

As mentioned above, detection of the T_g in a highly filled (does not contribute to change in heat capacity) or cross-linked (restricts polymer chain motion) polymer via DSC is difficult. As can be seen from Figure 5.3 the entire glass transition is easily observed from the DMA even in a highly cross-linked thermoset composite with high fiber loading (>60 wt.%). Unlike DSC, which measures the change in heat flow, the DMA measures changes in strength and energy loss. The magnitude of these response changes is much larger and therefore easily detected, making DMA the most suitable thermal analysis technique for measuring the T_g of composites.

Other useful DMA experiments for composites include stress relaxation and creep experiments. In a stress relaxation experiment, the sample is strained to a fixed displacement at a constant temperature and the decreasing change in stress is recorded. Stress relaxation can affect the warpage of a composite. In a creep experiment, the sample is subjected to a constant stress and temperature and the change in strain is recorded. Creep is important for understanding how a composite component will deform overtime in an application where it is under a constant load.

DMA instruments can be equipped with chillers to widen their testing temperature ranges for polymers with sub-ambient T_g's or for evaluating composites for low-temperature applications. Other DMA accessories and fixtures are available for testing in shear, compression, and tension (fibers/films) modes as well as humid and liquid environments.

THERMOMECHANICAL ANALYSIS

Common Uses in Composite Materials Characterization: coefficient of thermal expansion, glass transition temperature, melting point, and softening point

Understanding the dimensional stability of composite materials is crucial to their incorporation in multicomponent systems for any industry. Thermomechanical analysis (TMA) is a thermal analysis technique that measures the change in a sample's dimensions as a function of temperature, time, and force. In the field of composites, TMA is most frequently used to measure the coefficient of linear thermal expansion (CLTE) overprocessing or operating temperature ranges. Mismatch of CLTEs in joined or layered systems or even within a single composite article can lead to failure of the final component.

To run a TMA experiment on a fiber-reinforced polymer, the sample must have smooth, flat, parallel surfaces in contact with the stage on the bottom and a probe on top. As the sample is heated or cooled (typically in an inert atmosphere), the probe's movement records the expansion or contraction and the CLTE can be derived. Articles molded without surface texture are often suitable for measuring the CLTE in the z-direction, but composites typically have different CLTEs in at least two (or all three) directions. The CLTE in directions in which fibers are aligned will be dominated by the CLTE of the fiber. The CLTE in directions perpendicular or otherwise not in alignment with the fiber orientation will be dominated by the CLTE of the matrix. This is particularly important for unidirectional composites where all fibers are orientated in a single direction. Chopped fiber composites can have some degree of fiber orientation induced by flow during molding. Therefore, it is always best to measure the CLTE of composites in the x, y, and z directions, which usually requires sanding/polishing of the cut surfaces.

In addition to CLTE measurement, the T_g can be measured during a temperature ramp experiment. The TMA is more sensitive to T_g detection compared to DSC because it measures a different response, change in free volume rather than specific heat capacity. The TMA thermogram in Figure 5.4 shows the sample dimension change plotted against the temperature of a chopped carbon fiber-reinforced polyester measured in two directions under the same test conditions (5°C/min in nitrogen). The T_g can be measured from the z-direction curve (solid line) by taking the intersection of tangent lines below and above the slope shift. The entire glass transition is broad (approximately between 100°C and 175°C), which can be just as important as the assigned T_g value itself, which in this case is about 145°C. The CLTE in the matrix driven z-direction below and above the T_g is about 80 ppm/°C and 175 ppm/°C, respectively. The drastic difference in CLTE with respect to sampling direction is obvious. Carbon fiber has a slightly negative CLTE in the axial direction and when measured in the aligned x-direction, the CLTE of this composite below and above the T_g is about 2 ppm/°C and 4 ppm/°C, respectively.

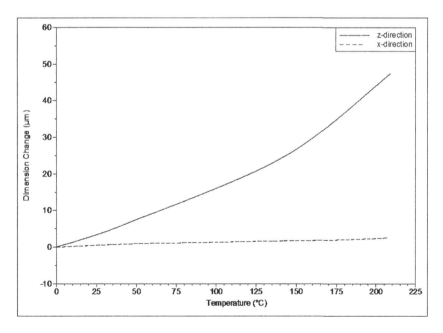

FIGURE 5.4 TMA (5°C/min in nitrogen) of a chopped carbon fiber polyester composite in both the z (solid line) and x (dashed line) directions; $T_g \approx 145$°C.

TMAs can be equipped with different probe and stage types to measure more than just CLTE and T_g of polymers and composites. A three-point bending fixture can be used to measure the heat deflection temperature, which is a useful property for material selection in long-term heated applications. Penetration probes can measure softening and melting points of thermoplastics and tension fixtures can measure stress/strain properties of fibers and films.

COMMON SPECTROSCOPIC ANALYSES

FOURIER TRANSFORM INFRARED SPECTROSCOPY

Common Uses in Composite Materials Characterization: chemical functional group identification, verification of materials composition, detection of contaminants, and degradation measurements.

FTIR is a common characterization and testing method found in composite laboratories that utilize the detection of representative chemical bond vibrations to identify and quantify chemical functional groups. It is typically considered a nondestructive analytical technique; however, the introduction of the sample into the spectrometer always dictates the proper size and larger solid composite samples must be modified (trimmed) appropriately.

Infrared radiation (near-, mid-, or far-infrared depending on the area of interest) is focused on the sample to measure the absorption intensities over a range of wavelengths. The wavelengths absorbed by the composite are characteristic of the vibrational (bending and stretching) frequencies of functional groups present in the sample. Absorption intensity at a characteristic wavelength indicates the concentration of the chemical group responsible for the absorption. This allows for quantitative analysis of chemical species. In composite applications, however, care must be used to quantify unique, characteristic peaks of the compound(s) of interest, being wary that oftentimes polymer blends, unknown additives, starting material by-products, processing aids, etc. may exhibit confounding absorptions in the region of interest and lead to inaccurate or misleading quantification.

FTIR is more commonly used for qualitative analysis. Functional groups are readily identified in the "group frequency region" from 3,600 to 1,200 cm⁻¹ and many references for assigning peaks in this region are available. Peaks in the "fingerprint region," typically from 1,200 to 600 cm⁻¹, are used mainly for comparison with a reference sample. Small differences in molecular structure lead to unique, significant changes in the appearance and distribution of absorption peaks in this region. Consequently, a close match between two spectra in the fingerprint region (as well as others) typically establishes solid evidence for the identity of the compounds.

FTIR is frequently used in composite laboratories for a variety of purposes. However, the most common uses are briefly outlined below.

Verification of Materials Composition: FTIR analysis is readily used to assess individual material components that constitute the composite. Since composites may consist of many different additives, resins, fillers, and fibers sourced from different suppliers, incoming quality control and regular monitoring of starting materials can reduce potential post-fabrication failures if discovered early, i.e., before the manufacturing process.

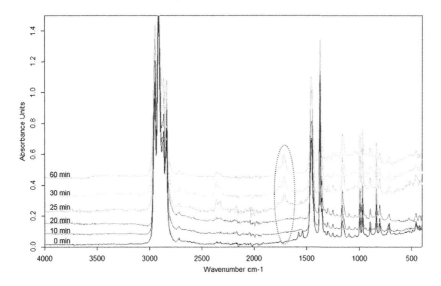

FIGURE 5.5 FTIR spectra of a typical polypropylene matrix with prolonged 200°C exposure. Thermal oxidation is indicated by the increasing $\nu C=O$ peak as circled.

Detection of Contaminants: The presence of minute impurities on composite surfaces does not typically lead to significant properties degradation. However, if two or more composite components are bonded together by adhesive, for example, such contamination could severely affect the strength and durability of the adjoining bonds. Hydrocarbon surface contamination such as oils from handling or machinery will also have detrimental effects on painted and coated composite surfaces.

Contamination within the composite structure may create and/or act as potential defect sites that can lead to a loss of strength. It can also lead to the formation of pits and blisters during molding, painting, or any other processing occurring at elevated temperatures.

Measurement of Thermal/Ultraviolet/Chemical Degradation: Hydrocarbon polymer resins, additives, and processing aids are usually very susceptible to degradation by heat, UV, and chemical attack. Thermal damage, for example, may oxidize resin components in composite materials which can lead to loss of strength and/or discoloration (when perceptible). In Figure 5.5, FTIR spectroscopy is used to identify and measure oxidative degradation of a polypropylene matrix when exposed to prolonged 200°C exposure, indicated by the growth of the $\nu C=O$ peak near 1,700 cm^{-1} over time.

Energy-Dispersive X-Ray Spectroscopy

Common Uses in Composite Materials Characterization: elemental analysis, verification of materials composition, and detection of contaminants.

The energy-dispersive X-ray spectrometer (EDS or EDX) is a powerful instrument for qualitative, quantitative, and (more commonly) semi-quantitative X-ray microanalysis of composites used in conjunction with scanning electron microscopy (SEM, see below). Simultaneous SEM and EDS are advantageous in that allow for chemical analysis of features directly observed in the microscope, making it ideal for spot failure/defect analyses of composite materials as long as a few very important factors are considered.

X-rays emitted from a specimen bombarded with a finely focused beam of electrons from the SEM can be used to characterize the elemental composition of the analysis volume, which is typically 500–3,000 nm depending on the accelerating voltage. Atomic core–shell electrons from the sample are excited to higher energy states by electrons from the SEM, leaving electron holes that are quickly filled by higher energy outer-shell electrons. The energy difference is released as an X-ray that has a characteristic spectrum based on the atomic structure of the element. Peak positions corresponding to electron binding energies identify the element, whereas the intensity of the signal(s) corresponds to its concentration. A typical EDS spectrum for a glass fiber thermoplastic composite is shown in Figure 5.6.

The elemental detection limits in EDS are highly dependent on the surface smoothness of the sample, with 2,000–5,000 ppm or 0.2–0.5 wt.% generally accepted as the lower limit for polished and stable bulk samples [1]. However, the lower limit of detection for many composite materials, especially traditional fiber-reinforced polymer composites, is generally much higher, and quantitative results are less precise as explained further below.

While the focused electron probe of the SEM is largely nondestructive for hard materials, many polymeric materials employed as composite matrices are vulnerable to decomposition by electron-induced reactions [2]. Thus the intensity of the electron beam must be cautiously managed especially if the original sample is subsequently reexamined by optical microscopy or other techniques. The possibility of chemical degradation of polymeric species by energetic electrons must also be carefully considered when interpreting semi-quantitative results. As a general rule, the accelerating voltage and beam current should be kept as low as possible to prevent sample damage and erroneous results.

Furthermore, semi-quantitative analysis of rough surfaces (such as composite fracture surfaces) is a great challenge due to the geometric effects that arise from remote, inconsistent scattering of electrons and absorbance of X-rays for rough

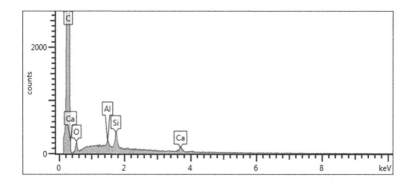

FIGURE 5.6 EDS spectrum of a glass fiber-reinforced thermoplastic composite.

surfaces compared to flat, homogenous bulk samples. Particles and fillers embedded in a composite matrix or the presence of voids near a smooth surface can have comparable scattering and absorbing effects that make reliable semi-quantification of composites difficult in highly filled systems [2]. Finally, highly filled composites are typically not homogeneous on the scale of the interaction volume at different locations. Thus it is not uncommon that semi-quantitative elemental analysis by EDS can vary at different positions on the same sample under the same analysis conditions. This is pointed out as care should be exercised when interpreting semi-quantitative results as EDS often only provides a "rough estimate" of the elemental composition of the composite material.

In contrast, EDS is highly useful for composite qualitative microanalysis, where the elements in a sample are identified quickly without the need to determine accurate abundances. For example, unanticipated elements in the EDS spectrum may point to possible contamination in defect or failure analyses, or the presence or absence of known elemental peaks can provide instant feedback to whether the correct materials or additives were employed during fabrication of the composite.

Typically the researcher must gather adequate spectrometer counts to ensure that a minimum peak size of three times the background is attained for accurate identification [2]. The analysis of composite samples requires special attention because there is always a risk of polymeric degradation with prolonged electron beam exposure that must be considered. However, polymeric degradation mainly leads to oxidative byproducts and even moderate oxidation is not expected to mask other elemental signals.

X-RAY FLUORESCENCE SPECTROSCOPY

Common Uses in Composite Materials Characterization: elemental analysis, verification of materials composition, and detection of contaminants.

X-ray fluorescence spectroscopy (XRF) is an elemental analysis method very similar to EDS. However, these methods differ in that XRF utilizes a primary X-ray beam (instead of a primary electron beam in EDS) that excites the sample surface. The X-rays generate holes in the inner electron shells that are filled by higher energy, outer shell electrons, leading to the emission of X-ray fluorescence radiation characteristic of the elements present in the sample.

XRF is generally a less destructive technique than EDS, as the X-rays used for the analysis seldom damage even sensitive samples. This technique is, therefore, more advantageous than EDS for the elemental analysis of composites with sensitive polymer matrices.

The depth of information gained from XRF is also typically higher than EDS because the area of X-ray irradiation is larger. This results in a more robust representation of elements present in composite samples as opposed to EDS. XRF is also more sensitive with detection sensitivity at the low ppm level. In addition, XRF gives true quantitative results for elemental abundances since it is calibrated with certified standards, unlike semi-quantitative (or standardless quantitative) analysis in EDS, in which the spectra are compared with data collected from standards stored within the system software.

Finally, X-ray irradiation and fluorescence X-ray detection can be carried out in air without the need for vacuum (as in EDS). This has led to the development of

handheld XRF analyzers which are capable of rapid elemental analyses during composite manufacturing or quality assurance on the factory floor.

Despite the above-mentioned advantages of XRF, SEM/EDS is popularly used
for micro-elemental analysis because it simultaneously provides SEM and elemental
distribution images (see more below).

X-Ray Photoelectron Spectroscopy

*Common Uses in Composite Materials Characterization: surface (nm) elemental analy-
sis, determination of elemental/chemical states, and depth profiling elemental analysis.*

X-ray photoelectron spectroscopy (XPS), also known as electron spectroscopy for
chemical analysis, analyzes the top 1–10 nm of a sample surface and is thus a true
surface chemistry measurement technique. XPS measures both the chemical composition and chemical bonding states of the elements.

A high energy X-ray source irradiates the solid sample surface and the kinetic
energies (or of the emitted electrons are measured and correlated to the shell of the
element from which the electron was ejected (e.g., 1s, 2s, 2p, etc.). These kinetic
energies are given as binding energies. Quantification of elements is determined by
binding energies and the intensities of the photoelectron peaks. Unlike EDS and XRF
mentioned above, XPS gives elemental information from the very top atomic layers
of the sample and is particularly useful in detecting surface contaminants and identifying the onset of substrate degradation from external chemical and environmental
exposures, which typically affect the surface of the sample preferentially before the
bulk material. Alternatively, additive compounds that are expected to bloom to the
composite surface, such as internal mold release agents or UV stabilizers, can be
confirmed readily by XPS analysis.

Since analysis by XPS is extremely surface sensitive, the composite researcher
should take great care not to introduce "artificial" contamination through excess
manual handling, cleaning, etc., as false peaks may arise that would not be nor-
mally expected. For example, the presence of sodium *and* chlorine (as NaCl) is
indicative of sample handling without gloves. Handling the sample without gloves
can also contribute to a strong carbon C1s signal that may obscure the relative
abundances of other elements in a survey scan and should, therefore, be avoided
with care.

XPS is considered a nondestructive analytical technique, but in rare cases, certain
polymers may be sensitive to prolonged X-ray exposure. Therefore, care should be
used to minimize analysis times when appropriate. In contrast, XPS depth profiling
experiments are purposely destructive as the elements underneath the surface are
probed by first removing surface atoms by ion sputtering.

A powerful capability of XPS analysis is that spectral peaks in the "core regions"
(binding energies >30 eV) can be fitted computationally with component peaks to
separate the photoemission signal from distinct elemental and/or chemical states.
The binding energy or position of a component peak allows for the assignment
to a specific chemical environment and concentrations of the chemical states are
determined by measuring the component peak areas. However, caution should
be observed when peak-fitting as the challenge arises to make mathematical and

physical models agree [3]. An example of peak-fitting for a carbon (C1s) peak is shown below in Figure 5.7. In this particular case, the model was chosen based on the chemistry of the analyzed sample and literature references, being also mindful of the confounding presence of adventitious carbon on the surface. Peak-fitting in this example shows the presence and relative abundances of C1s C-sp^3, CO, and COO. Component assignments and concentrations for carbon and oxygen especially can be extremely helpful in composites research, as most degradative polymer pathways are oxidative in nature and significant carbon oxidation on the surface can indicate the beginnings of substrate failures.

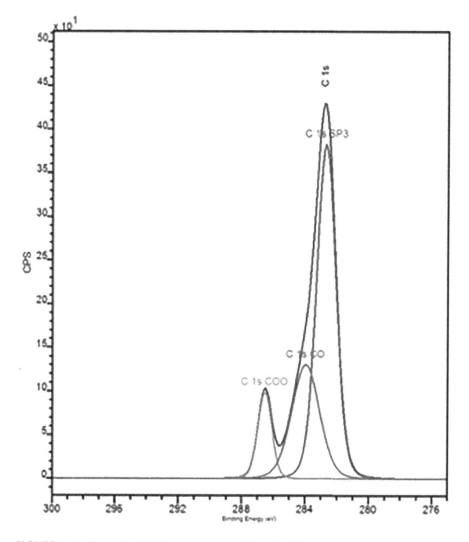

FIGURE 5.7 Example of XPS software peak-fitting for C1S photoelectron peaks showing the presence of C-sp^3, CO, and COO.

SCANNING ELECTRON MICROSCOPE AND
ANALYSIS OF COMPOSITE MATERIALS

Common Uses in Composite Materials Characterization: failure/defect analysis, fiber–matrix interactions, and composition and topography investigations.

The scanning electron microscope (SEM) has become a very versatile and powerful instrument available for the examination and analysis of composite materials. The utility of the SEM stems from its capability to render highly detailed 3-D images of the surfaces of a wide range of organic and inorganic solid materials. The development of variable-pressure SEMs allows for the imaging of insulating or biological materials without the need for conductive coatings. While commercial SEMs usually state instrument resolutions < 10 nm, many composite researchers will operate routinely from the μm to nm scales as different information is often collected at different magnifications (typically 10–10,000×). Very low magnifications are employed to complement information gathered with an optical (light) microscope, which is still highly used today in composites analysis. However, the SEM has a very large depth of field, or vertical height range over which the sample still appears to be effective in focus, that is vastly superior to optical microscopes and accounts for its ability to readily deliver 3-D and easily interpretable images.

The sample area to be examined in the SEM is irradiated with a finely focused electron beam that rasters across the surface at the decided magnification. The interaction of the beam and the material (at the "interaction volume") produces detectable emission signals from backscattered electrons (BSEs), secondary electrons (SEs), and characteristic X-rays (see EDS above) among others. Detection of backscattered or secondary electrons typically forms the SEM image (micrograph) and provides different information about the sample as described further below. Both BSE and SE SEM images are frequently used in the analysis of composite materials; however, it is very important to understand the differences between the two.

Interaction of the primary electron beam with the various atomic nuclei of the sample produces high-energy BSEs (originating from the primary electron beam itself) through multiple elastic scattering events. BSEs typically originate from deeper regions (>1 μm) within the sample and thus contain "sub-surface" information which can be controlled to some extent by the accelerating voltage [2].

Atoms of elements with high atomic numbers (Z) scatter electrons more easily than those with low atomic numbers and, therefore, produce a greater detectable BSE signal. This increase in signal with increasing Z provides a mechanism for generating a compositional contrast in the BSE image. For example, atoms of silicon (Z = 14) will scatter more electrons toward the detector than carbon (Z = 6) so that silicon-rich areas of a specimen will appear brighter than carbon areas in the SEM image. The effect of this atomic number contrast is strongest below Z = 50 [2].

BSE contrast dependence on the atomic number helps differentiate the different phases of a material, providing imaging that carries information on the sample's composition and topography. The contrast in BSE images can also provide valuable information on crystal orientation and magnetic properties [4]. However, the BSE signal also reduces spatial resolution (compared to the SE signal described below) due to lateral scattering, resulting in a decreased capability to resolve fine features of the sample.

Low-energy SE emission is a result of inelastic scattering of the primary electron beam and the loosely bound outer shell electrons of the sample. SEs originate from the sample at the surface or near-surface (~20 nm) and SE SEM images are thus more surface sensitive, showing a high amount of surface topography and texture (see Figure 5.8). SE images tend to be brighter at the edges of sharp features and darker when flat (perpendicular to the primary electron beam), thus giving well-defined 3-D micrographs.

The high magnification, high-resolution imaging of SEM analysis is commonly used to investigate composite fracture surfaces, including fiber failure, fiber surfaces, fiber–matrix interfacial adhesion, fiber uniformity, etc., as shown in Figure 5.8. It is also utilized for investigating surface flaws and for the identification of contaminants or unknown particles and the interactions among component materials. In conjunction with EDS, the SEM becomes a powerful and highly useful resource for the study of composites.

However, the investigator must endeavor to image as many areas and/or samples as possible before making generalized statements concerning composite behavior or properties, primarily for the reason that a single SEM image commonly represents only a very small area of examination and may not be representative of the entire composite as a whole. Acquiring multiple images also reduces the complication of image interpretation, as multiple factors such as electron beam interaction, atomic number differences, and topography will likely change from image to image.

Potential consequences of using the SE or BSE detector when imaging a composite sample are shown in Figure 5.9. In SE mode, a single glass fiber appears just above the surface of the thermoplastic composite, still covered in a very thin layer

FIGURE 5.8 Fracture surface of highly filled glass thermoset composite in SE mode showing detailed topography and surface texture.

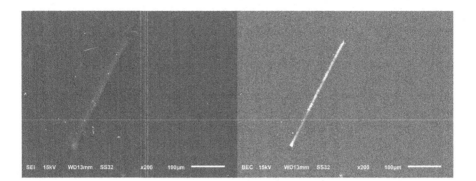

FIGURE 5.9 Single glass fiber above the surface of a thermoplastic matrix seen in both SE mode (L) and BSE mode (R). The fiber appears to be exposed and below the surface in BSE mode due to the depth of the interaction volume at an accelerating voltage of 15 keV.

of polymer matrix. In BSE mode, however, the fiber appears as if it is *completely exposed and just below* the surface. The SE image shows the topography of the fiber protruding above the flat surface exhibiting fine 3-D surface detail; BSE signals penetrate deeper into the sample and the compositional contrast, coupled with the loss of topography, makes the fiber appear below the surface. The fine line transverse to the fiber in the SE image is a tooling mark from the polished metal mold, which is not apparent in the BSE image indicating it is indeed a surface artifact. Thus at 15 keV accelerating voltage the BSE signal is unequivocally not a surface-sensitive signal yet still gives valuable information (albeit indirectly) when considered with the SE image.

GAS CHROMATOGRAPHY–MASS SPECTROMETRY

Common Uses in Composite Materials Characterization: chemical identification and analysis of volatile organic compounds (VOCs).

Gas chromatography–mass spectrometry (GC–MS) is a synergistic coupling of two very powerful microanalytical techniques that allow for the separation of a chemical mixture and the identification of its components. GC–MS is routinely used in many applications where rapid, high sensitivity, and accurate analysis of volatile compounds is necessary.

GC–MS has the capacity to resolve very complex mixtures of hundreds of compounds, where a vaporized gas, liquid, or solid sample is injected and propelled through a capillary column coated with a stationary phase by an inert carrier gas such as helium or nitrogen. The components elute from the column at different times (called retention times) depending on adsorptive (or partitioning) interactions and are usually predicted by the compound's polarity and boiling point. After exiting the GC column, the separated components are ionized and fragmented by a chemical ionization or electron impact source. These ionized fragments are subsequently accelerated through the mass analyzer where their mass-to-charge (*m/z*) ratios are measured. Importantly, compound fragmentation patterns are unique and reproducible such that extensive fragmentation "fingerprint" libraries have been compiled to allow the researcher to quickly and accurately identify unknown substances within

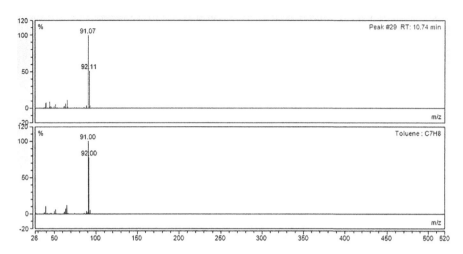

FIGURE 5.10 Electron ionization mass spectrum for toluene (C_7H_8) reference standard (a) compared to NIST library match (b).

reason (Figure 5.10). There are currently over 2 million mass spectra for over 350,000 compounds currently included in the National Institute of Standards and Technology (NIST) Mass Spectral Library [5]. However, it must be cautioned that the *absolute* identification of unknowns can only be accomplished by direct comparison with reference compounds injected in the same instrument under the same conditions, ensuring to match both the retention times and mass spectra for each.

GC–MS is essential in the quantification of volatile and semi-volatile organic compounds (VOCs and SVOCs) in composite materials. Volatile compounds in composites can be attributed to unreacted monomer, processing aids, decomposition products, contaminants, etc. These volatiles are also responsible for odors that many people often find offensive, especially when composite materials are used in confined spaces with limited airflow (e.g., interior transportation compartments and building interiors).

Static GC headspace analysis, a widely used sampling technique, analyzes the vapor phase that is in thermodynamic equilibrium with the sample in a closed system (typically a vial). This method is limited due to the preference of highly volatile compounds to partition into the headspace volume, potentially under-representing the presence of higher boiling point VOCs in the sample.

Thermal desorption (Purge and Trap) GC–MS resolves these issues by heating the composite sample to specified conditions under a purge of gas flow, where volatiles are adsorbed onto a separate, porous polymer adsorptive or "trap." This type of dynamic headspace sampling prevents an equilibrium between volatiles and the heated sample from forming, allowing higher boiling point compounds to vaporize freely and thus resulting in a more accurate determination of the VOCs in the composite sample. Sensitivity of thermal desorption GC–MS is typically 1,000 times higher (ppm to ppb range) than traditional static headspace analysis since theoretically all of the volatiles from the sample are removed from the matrix onto the trap in the absence of an established equilibrium [6].

No matter what sampling technique is employed, considerable effort must be used to ensure that the composite sample to be analyzed (typically mg quantities) represents the consistency of the entire composite material (i.e., g or kg quantities) *as much as possible*. As mentioned previously, the amount of fiber (or conversely the amount of matrix) in the analyzed sample affects thermal events in DSC and TGA. Similarly, varying amounts of matrix materials dramatically affect the VOCs and SVOCs detected by GC–MS and can lead to irreproducible or misleading results. Therefore, analyzing repeat samples of the same geometry, surface roughness, fiber and matrix content, mass, etc., can mitigate these effects. Inhomogeneous resin-rich (resulting in higher detection of VOCs) or fiber-rich areas (resulting in lower detection of VOCs) should be avoided if those areas do not faithfully represent the entirety of the composite. Multiple replicates sampled from different areas of the composite should be tested and attributed as a best practice.

OTHER TESTING

As mentioned previously, there are too many characterization techniques in the field of composites to properly discuss them all in detail in this book where the scope is intentionally broad. Interest in any of the topics above warrants a more in-depth literature review by the researcher. In this last section, more key material properties are briefly highlighted in an effort to simply make the reader aware of their existence for potential further research.

MECHANICAL PROPERTIES

Typically, the most important mechanical properties in the composite industry are the tensile and flexural strengths and moduli. Tensile properties are collected by loading a flat test specimen (often machined into a "dog bone" shape) of specified dimensions into a universal testing machine that applies an increased tensile load until failure and records a stress–strain curve. In addition to tensile strength, other properties such as tensile modulus (Young's modulus, measure of stiffness), Poisson's ratio (describes the shape change in perpendicular directions), and elongation at break (measure of ductility) are recorded or derived. A universal testing machine can be loaded with other fixtures to collect compression or flexural (three-point bending) data. Dividing these strengths/moduli by the composite's density allows one to derive the specific strength/moduli to determine attributes like strength-to-weight ratio, which is useful when comparing composites to other materials like metal alloys.

Other important mechanical properties include hardness (resistance to indentation or abrasion), fatigue (repeated loading), and toughness (important for energy absorption and impact resistance). Interfacial shear strength can be measured by single fiber pullout testing in a matrix and provides important information about the adhesion and load transfer between the fiber and the matrix. All of the mechanical properties described in this section can be used by engineers to model composite systems for component design.

Rheology

Rheology is the study of flow and is useful for not only designing composite process techniques but also for understanding and controlling the impregnation of fiber with the matrix (also known as fiber wet-out). There are many different styles of rheometers on the market. For example, a parallel plate model sandwiches a liquid sample between two plates and applies a rotational twisting motion (shear). A temperature ramp is then applied to observe, for example, the change in the polymer's viscosity (resistance to flow) as a function of temperature and time. The viscosity of a liquid polymer matrix is a key factor taken into consideration when fabricating composites as fiber wet-out is critical to achieving load transfer from the matrix to the fiber. Simple viscometers provide a quick measurement of a material's viscosity at a single temperature and can be used to ensure proper fiber wet-out during fabrication.

Miscellaneous

Optical microscopy is widely used in the composites industry and often as a tool for failure analysis. By cross-sectioning the defect area (could appear as a crack, pit, depression, raised area, etc. on the surface) of a composite, one can look for the root cause of the flaw in the bulk, which could perhaps be a foreign object, a void, delamination between layers, poor fiber wet-out, etc.

The surface energy of a composite can be measured in several different ways (water contact angle, dyne pens, etc.) and provides helpful information about how easily the composite will bond to coatings, paint, or adhesive.

Weather-ability tests employ UV, humidity, salt spray, etc., chambers to simulate real-world conditions on composite samples and typically measure some sort of response before and after conditioning such as moisture uptake, color change, or mechanical property loss.

REFERENCES

1. Wolfgong, W.J. 2016. Chemical analysis techniques for failure analysis: Part 1, common instrumental methods. In *Handbook of Materials Failure Analysis with Case Studies from the Aerospace and Automotive Industries*, ed. A.S.H. Makhlouf, and M. Aliofkhazraei, 279–307. Waltham, MA: Elsevier.
2. Goldstein J., Newbury, D., Joy, D., Lyman, C., Echlin, P. Lifshin, E., Sawyer, L., and J. Michael. 2003. *Scanning Electron Microscopy and X-Ray Microanalysis*. New York: Springer.
3. Major, G.H., Fairley, N., Sherwood, P.M.A., Linford, M.R., Terry, J., Fernandez, V., and Artyushkova, K. 2020. Practical guide for curve fitting in x-ray photoelectron spectroscopy. *J. Vacuum. Sci. Tech. A* 38: 061203 1-061203-22.
4. Robinson, V.N. 1980. Imaging with backscattered electrons in a scanning electron microscope. *Scanning* 3: 15–26.
5. NIST Standard Reference Database 1A. 2021. https://www.nist.gov/srd/nist-standard-reference-database-1a.
6. Sparkman, D.O., Penton, Z.E., and F.G. Kitson. 2011. *Gas Chromatography and Mass Spectroscopy - A Practical Guide* Second Ed., Burlington, MA: Elsevier.

6 Innovations in Automotive Composites through the Years
Strength and Gaps

Probir Guha

CONTENTS

Over the years, automotive parts made out of steel have held a significant cost advantage over any other material including aluminum, magnesium, and composites. In this chapter, we will try to understand what is it that drives these costs and innovations in composites that have consistently worked to diminish this value gap between steel/aluminum/magnesium and composites.

No matter the industry – cost AND performance drive customer decisions. From a product perspective material, product design and manufacturing process drove costs. Companies in different industry segments had to understand and drive innovations in these three areas to remain competitive.

While I was driven by the needs of the automotive industry in my development role, what I observed in my sales and market development role is that innovations in automotive very often would be just as important for marine or farm equipment or heavy truck or entry door applications.

In this chapter, I have enumerated some of process innovations that have been key to composites in the past 50 years. And will also try to point out some areas that could merit further significant improvement in the next 50 years.

To fully appreciate the important role that manufacturing processes play in the growth of composites applications, it is important to understand the predominant

DOI: 10.1201/9781003161738-6

structural material of choice – stamped steel. And the manufacturing process used for steel has been stamping. A typical steel stamping manufacturing process flow chart is shown in Figure 6.1:

The steel and later aluminum stamping process is defined by the following characteristics:

- Requires progressive dies to avoid spring back/improve dimensional stability
- This leads to high tooling/die cost
- Development of stamping process software to understand stamped steel "wrinkling" issues and spring back during the stamping process
- Very fast cycle time – high throughput
- Most unused metal is recycled
- Over the years, the metal stamping material and process have become highly standardized and automated.

A typical SMC compression molding manufacturing process flow chart is shown in Figure 6.2. While some of what is depicted are typical of body panel applications, the process steps are generic of the composite compression molding process. These steps have to be viewed in comparison to the metal stamping process steps shown in Figure 6.1. The overall product quality, non-value-added steps included in the sequence, and cost can be evaluated by studying this flow diagram.

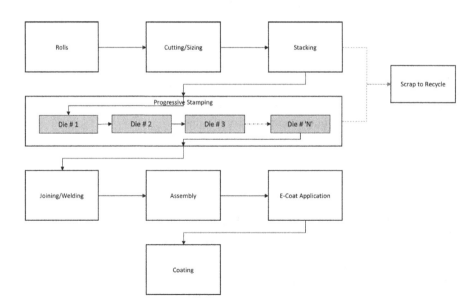

FIGURE 6.1 Process flow for a metal stamped part. (The dotted flow line shows non-value-added process stream.)

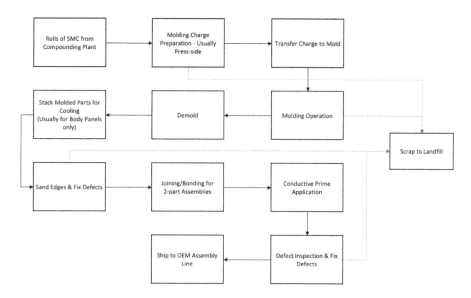

FIGURE 6.2 Process flow for an SMC compression molded part. (The dotted flow line shows non-value-added process stream.)

The SMC compression molding process is defined by the following characteristics:

- Does NOT require progressive dies as is typical for metal stamping.
- This leads to lower tooling/die cost in comparison to stamped metal.
- While dimensional variance due to "spring back" is not prevalent in molded SMC parts, there is a need for a theoretical understanding of the effect on dimensional variances directional chopped fiber distribution in the x–y plane of a molded surface.
- Mold cycle time is substantially higher than metal stamping, and this leads to a higher number of tools/molds required for a specific production volume.
  - A typical comparison of tooling costs between composite molding process and metal stamping is shown in Table 6.1.
- Variability in molding charge pattern and placement can lead to inconsistencies in the part.
- The SMC molding process is defined by both the chemical kinetics and the flow rheology of the SMC during the molding process. Variability can lead to defects in a molded part.

In recent times, the resin transfer molding (RTM) process is making significant advances in high volume products as well. Therefore, it behooves us to review the RTM process. The RTM process has been very successful when used to make larger parts with lower volume requirements. However with the advent of innovations in the RTM process, mold design technologies, innovations in preforms with higher permeability, this process has been moving into the mainstream of automotive applications. While some of what is depicted are typical of body panel applications, the process steps are generic of the RTM process and Figure 6.3 depicts a typical RTM molding process.

TABLE 6.1
Total Cost Comparison for a Generic Part

		Steel	Aluminum	GF-SMC (Standard Density)	GF-SMC (Low Density)	GF-RTM (Continuous Fiber)	CF-Autoclave	CF-SMC
Weight	lbs	10	6	7.5	6.5	5	3.5	5.5
Material	$/lbs	$0.50	$2.00	$1.00	$2.00	$5.00	$25.00	$8.00
	$/part	$5.00	$12.00	$7.50	$13.00	$25.00	$87.50	$44.00
Volume	Number/year				50,000			
Cycle time	Seconds/part	15	15	120	120	120	3600	120
Direct labor	$/part	$0.08	$0.08	$1.33	$1.33	$1.33	$40.00	$1.33
Mold	$/part	$0.85	$1.10	$0.50	$3.75	$0.50	$3.75	$0.50
Total cost	$/part	$5.93	$13.18	$9.33	$18.08	$26.83	$131.25	$45.83
$ increase per lbs saved (vs. Steel)		-	$1.81	$1.36	$3.47	$4.18	$19.28	$8.87

Numbers used are generic and should be used for comparison and educational purposes only.

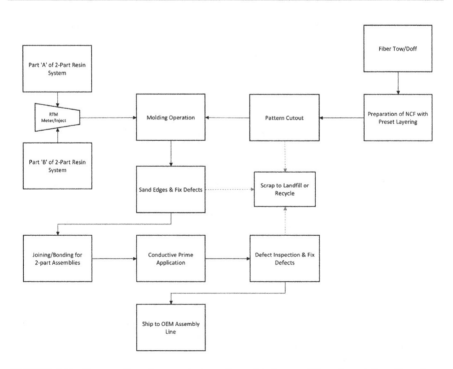

FIGURE 6.3 Process flow for a resin transfer molded part. (The dotted flow line shows non-value-added process stream.)

The RTM process is defined by the following characteristics and some of its strengths and weaknesses are similar to the SMC process:

- Does NOT require progressive dies as is typical for metal stamping.
- This leads to lower tooling/die cost in comparison.
- While dimensional variance due to "spring back" is not prevalent in molded RTM parts, there is a need for a theoretical understanding of the effect on dimensional variances directional chopped fiber distribution in the x–y plane of a molded surface.
- Mold cycle time is substantially higher than metal stamping, and this leads to a higher number of tools/molds required for a specific production volume.
  - Over the past several years, innovations in processing equipment and resin chemistry have reduced the cycle time gap between metal stamping and RTM and brought RTM more in line with SMC molding cycle times.

A typical comparison of tooling costs between the composite molding process and metal stamping is shown in Table 6.1. In this table, we have used typical assumptions for weight reduction, costs associated with material, process, and molds to make the part. Numbers have not been adjusted for poor quality and defects. This database is meant to provide a quick view of the strengths and weaknesses of various materials.

In terms of cost alone stamped steel components have a significant advantage when compared to other materials. And we can also see that carbon fiber composites have the lead on weight reduction capability – at a cost. Figure 6.4 utilizes the data presented in Table 6.1 to show the relationship between the cost of weight reduction and actual achievable weight reduction. As we achieve more weight reduction, the cost keeps going up. This cost is reported as the cost increase due to weight reduction for each pound of weight reduced. So in redesigning a stamped steel part in stamped aluminum will have a lower cost increase and a smaller percent weight reduction when compared to what can be achieved with a carbon fiber composite. In the materials selected in this database, the cost of weight reduction ranges from $1.81 for stamped aluminum to $19.28 for autoclave molded carbon fiber composites. The user will factor this cost increase against the benefits accrued from specific applications when deciding which material technology to use for the part. Aerospace applications derive a much higher value from weight reduction and have gradually implemented increasing amounts of carbon fiber composites in their components. In comparison, automotive has been more cautious in adopting light-weighting technologies. Over the past several years, automotive applications have moved toward an acceptance level of $2 for the cost of weight reduction. Hence the increased acceptance of aluminum in automotive applications. The current emphasis on automotive electrification has increased the need for further weight reduction. Will this $2 cost of weight reduction target be driven up a notch? Recent advances in composites are filling that gap of cost and weight reduction that exists today between aluminum and conventional carbon fiber composites.

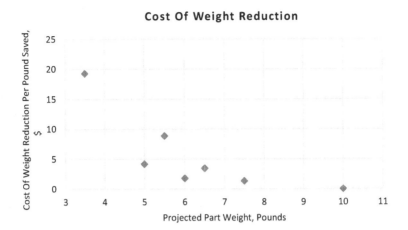

FIGURE 6.4 Cost of weight reduction. (The numbers used are generic and should be used for comparison and educational purposes only.)

For composites the effort to be the material of choice for weight reduction, especially in automotive, has primarily focused in the following areas for innovation:

- Materials and process innovations that go to the core of the weight reduction and productivity challenges
- And product differentiation and migrating into applications that have been traditionally in steel

We can learn from some of the key the innovations over the past 50 years.

INNOVATIONS IN PRODUCTIVITY AND WEIGHT REDUCTION IMPROVEMENTS

Paint defects. Pits and Porosity. The scourge of the SMC industry.

Most automotive body panels in composites were compression molded using sheet molding compound (SMC). The molded part after going through secondary and joining operations would then go through multiple layers of coating that included a layer of conductive primer, and the color coats. Pits and porosity are tiny holes or craters on a molded composite part. These tiny "defects" would lead to amplified visual defects following the painting process.

A visual defect on a painted part is a cardinal sin in automotive!

All the cost benefits achieved through the lower cost of molds for composites in comparison to the cost of molds for the progressive stamping process used for steel were lost – and then some – in the effort at the final paint line at the Original Equipment Manufacturer (OEM) factory.

This added direct labor – COST

This slowed down throughput – COST

The paint defects challenge to the growth of SMC led to a series of key innovations over the years.

IN-MOLD COATING

Gencorp – pioneered the in-mold coating process [1] in 1990 to address the paint defect issues (Figure 6.5). This was considered to be a major automotive OEM development to apply a coating on the molded part in the mold itself. Hence the name "in-mold coating." The technology was designed to force a paintable *in situ* polymer at the end of a molding cycle into the molding defects left on the part due to molding process and material inability to provide a part without any surface blemishes or porosity as they are called. These surface defects would lead to significant appearance defects on the final top-coated part.

Defects. Rework. Cost

The process showed promise. One significant drawback was the increase in cycle time required to crack the mold open a tiny bit at the end of the molding cycle to inject the in-mold coating onto the surface of the molded part. Eventually, this drawback was eliminated by injecting the IMC under high pressure without "opening" the mold. A secondary issue was faced on the coating uniformity on the vertical surfaces of the molded part. Unless the press had superior parallelism, hydraulic control in the movement of the press in the horizontal direction during the IMC inject phase would cause untreated areas. However, by and large, this process had significant success in greatly diminishing surface defects with subsequent reduction in paint rework and scrap.

VACUUM MOLDING

In-mold coating with all its strengths had a weakness in vertical walls of a part, as has been stated above, and not very effective along the edges of a molded part. The Budd Company introduced an improved vacuum molding process in 1988 [2] and now has gone through several generations of improvements [3]. It is still in use in compression molded SMC applications to this day. Two major areas of success for

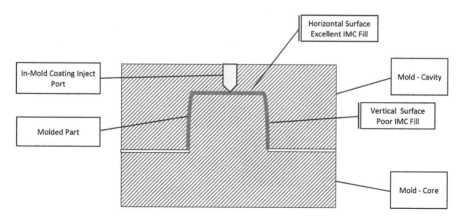

FIGURE 6.5 In-mold coating (IMC) process schematic.

SMC – body panels and the pickup box – would have faced significant challenges had this vacuum molding process not been used.

In Figure 6.6, I have shown some of the important aspects of the molding process as described in the patents referenced earlier.

Figure 6.6 depicts the three key steps in the mold closing sequence under vacuum. This is a depiction and while it emphasizes the important features it certainly is not to scale and lacks detail. The key features shown are as follows

- Mold is constructed with stops to protect the mold surface at full mold close position.
- Mold is constructed with a vacuum seal and vacuum line.
- Vacuum is actuated as soon as the vacuum seal is engaged.
- Vacuum is engaged typically at mold position of 3 inches off stops.
- Dwell time of vacuum from initiation to before the mold core touches the SMC charge is 3 seconds or more.
- Vacuum level achieved in the vacuum cavity has to be at least 26 inches of water (or more) before the mold core touches the SMC surface. Vacuum levels lower than 26 inches rapidly become ineffective toward eliminating defects.
- At full close, the mold is off the stops to ensure that the pressure is on the SMC.

Over the years on innumerable occasions, I have seen molding operations drifting away from the features described above which in turn led to defects in the paint due to surface porosity in the molded part or air entrapment in the molded part causing defects commonly referred to as "blisters."

Like I mentioned before, the success of SMC body panels and pickup box applications was in a large measure due to successful implementation of vacuum molding. The process described was first implemented in 1988 and then refined in 2014 and still used widely. A lot of what exists today was done with experimentation and trial and error, and there is certainly room for improvement through greater theoretical understanding of process, adding state of the art electronic controls, and data analysis. Air entrapment will always be a challenge in a molding process where

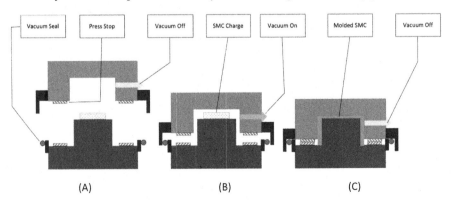

(A)   (B)   (C)

FIGURE 6.6 Key steps in the vacuum compression molding of SMC.

a low viscosity moiety transforms into a solid component during the process and controlling that aspect with vacuum can only lead to better products and manufacturing productivity – something to keep in mind as the traditional compression molding process morphs into overmolding and hybrid molding with continuous fibers.

Tough Class 'A' SMC and Low-Density SMC

In-mold coating was a technology to minimize the defects caused by surface porosity on a molded part after they had been caused, and vacuum molding addressed the root cause of the porosity defects and sought to eliminate their occurrence. However, both of these techniques focused on the defects on the primary surface of the part, not on the edge of the part. Two very consequential innovations that led at first an effective solution for molded part edge defects and then taking the ensuing material to a low-density form were published in patents and are described below. The first innovation was used in a product known as Tough Class A SMC or TCA [4], and its phenomenal success in reducing defects in top-coated parts is shown in Figure 6.7. The vertical axis is a measure of defects in the form of number of defects per a thousand vehicles – for a typical stamped steel body panel this defect rate was in the vicinity of 25 – so that was the target the SMC molder had to achieve. The data shows that TCA took the defect rate down to just above zero! The TCA innovation was a major milestone in the development of composites for body panel applications and became the cornerstone of several new innovations in the coming years.

A few years later, TCA was further enhanced by substituting some or all of the high-density filler typically used in SMC with hollow glass bubbles to form a low-density product. The chemical treatment given to the glass spheres helped the SMC to retain the mechanical properties of its standard density predecessor. This product was named TCA Lite and then further use of the treated glass hollow spheres led to TCA UltraLite [5].

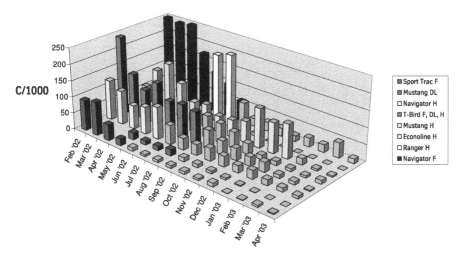

FIGURE 6.7 Defect reduction by using TCA-SMC in automotive body panels. (Courtesy CSP-Teijin.)

INDUCTION HEATING AND PREHEATING TO IMPROVE MOLDING CYCLE TIME

A significant gap between plastics – thermosets and thermoplastics – and stamped steel existed in the cycle time for this forming operation. And there were efforts both in terms of materials and process innovations to close this gap. In the 1990s, The Budd Company introduced a technology to preheat the SMC charge prior to the charge being transferred to the mold (Figure 6.8) [6]. This extra-mold heating phenomenon would reduce the heating load while the charge was in the mold affecting a reduction in cycle time. The technology was discussed in the published patent shown below.

A few years later, the "heating" mindset was extended to the rapid heating and cooling of molds as well [7]. This was introduced by Roctool and is still gaining acceptance in many industries in both thermoset and thermoplastic applications.

This technology described by Roctool merits further consideration and additional development in the molding of thermosets and thermoplastics. The preheating technology described in the Budd patent above was used in production, but over a period of time, the technique was "moth balled" due to the inability of the process to accurately and consistently control the preheat temperature of the molding charge.

A coming together of these two technologies described may help the user better control the preheat/induction heating concepts. Another important point to note is the mention of Curie Fillers in the Roctool patent. A Curie Filler is a material that in a magnetic induction field will heat up to a specific temperature at which it then loses its capability to respond to the magnetic field and induce further temperature increase. Therefore, quite possibly this type of filler could address the core temperature control issues faced by the Budd preheat technology.

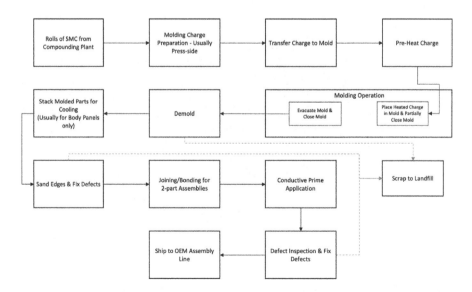

FIGURE 6.8   Process flow with preheated charge for SMC compression molding.

This type of filler has been claimed in various types of adhesives but has yet to be adequately pursued in the formulation of polymeric molding compounds. As long there exists a manufacturing throughput gap between stamped metal and molded plastic, this is phenomenon that needs to be studied further.

FAST CYCLE RTM

Over the years, RTM has been utilized in low volume applications. However, with the advent of "fast cure" epoxy and cost-effective continuous fiber technologies, RTM has become relevant for high-volume applications as well. This improvement was disclosed in a patent in 2019 [8].

In the RTM process, resin flows through the fiber preform has often been the limiting factor for minimizing cycle time. A lot of the applications were based on chopped fiber preforms and the "low-"permeability of these preforms would require high gel time and long cure resin systems. In recent times, several innovations have been catalysts for rapid cycle time in RTM and also in wet molding process. Even though I do not have a separate discussion on wet molding processes, a lot of the innovations in RTM would also be useful for wet molding processes.

The primary innovations that have paved the path for rapid cycle times in RTM processes are as follows:

- Special equipment for HPRTM (high-pressure RTM). This innovation was a derivative of the reaction injection molding process used in urethane applications.
- Multiple port injection technology as has been discussed in the patent above.
- Introduction of cost-effective continuous fiber composite technology using traditional textile fiber laying technology. This has been discussed a little further down in this chapter.

These are emerging developments and a lot more has yet to come as the engineering and scientific community finds a way to understand and define the interaction of preform permeability; ways to control permeability in different sections of the same preform; and resin cure kinetics and rheology.

The future of composites looks very promising with these recent advances in the RTM (and wet molding) process.

The next few examples of innovations discussed below will focus in the area of continuous fiber in composites. Over the years – both for thermoplastics and thermoset composites – the industry has opted to focus on the use of short or chopped fiber composites. We all know that the strength and stiffness in the composite are derived from the reinforcing fiber in the composite. We also know that when we go from a continuous fiber composite to a chopped fiber composite a very high percentage of the mechanical properties are degraded by as much as 75% or more. As the composites industry faces the task of competing with aluminum and magnesium for cost-effective weight reduction, increased use of continuous reinforcing fiber becomes very important.

EARLY USE OF CONTINUOUS FIBER IN AUTOMOTIVE APPLICATIONS

The first application discussed is for the energy attenuator, an energy management device for an automotive bumper. This application uses continuous fiberglass and a filament winding process for significant weight reduction compared to the prevalent steel energy absorbing system. While the application was a very significant engineering success, it was not able to meet the cost expectations of the automotive customer. In the 1970s even with the increasing energy costs at that time, the automotive market was not ready to pay a premium for weight reduction and better gas mileage in a vehicle. Maybe the innovation [9] was a bit ahead of the needs of the market. But it was a very clever use of continuous fiber and made its way into the current discussion.

Another early example of the effective use of continuous fiber from the 1970s and 1980s was the automotive leaf spring [10]. This was a very significant weight reduction over the prevalent steel leaf spring. And even though it saw limited acceptance in the early days, it does seem like the time for this concept has arrived and composite leaf springs are more in use today. The patent cited here [10] based the application on a filament winding process and the associated process cost at that time may have been the cause of its slow acceptance. In today's world, the market is using similar leaf springs made by compression molding prepregs and are now looking into products made by the RTM process.

Both these applications were clever to use of continuous fiber for significant weight reduction in applications deemed difficult in chopped or short fiber systems. Today we see the need for continuous fiber reemerging as the composites industry explores ways to distance itself from its aluminum and magnesium competition and start the journey into structural components in a vehicle.

THE NEW CONTINUOUS FIBER FOR COMPOSITES – LATTICE

New thoughts and solutions are entering the composites arena. That is an excellent development. I have mentioned earlier that the textile industry, more specifically Coats, has decided to bring its solutions to the world of composites. An innovation disclosed in 2020 [11] discusses this textile fiber handling technology for continuous fiber composites. Coats has a name for this family of embroidered products – Lattice.

Lattice is a development to keep an eye on – this has immense potential. This technology addresses the cost and performance gaps that the composites industry is looking to fill to pull ahead of aluminum and magnesium in the sector of "cost-effective weight reduction." So what do we stand to gain by using Lattice?

*   Use of continuous fiber compared to the more prevalent short and chopped fiber in high-volume applications
*   Can use dissimilar fibers on the same preform. So the user can have carbon where high stiffness is required and glass fiber elsewhere. This helps control cost

- The technology makes near-net-shape preform. Hence no waste – big gain in terms of cost-effectiveness
- The embroidery techniques associated with textiles allow the fabricator to build in "mistake proofing" features and assists for exact placement of the perform in a mold
- Smart composite sensors, actuators, and conductive paths can be embroidered into the Lattice product
- And the Lattice technology is compatible with a myriad of plastics processing methods

We have now gone through some of the author's thoughts on key innovations in composites in the past 50 years primarily specific to the automotive industry. We have seen how some of these have led to commercial success and others have met with limited commercial viability at the time the innovation was made. But we learn from our experiences and it is hoped that this paints a picture of what needs to be done now and in the future to meet current challenges faced by the composites industry. We then concluded with a discussion of a current foray into the composites world with solutions offered by the world of textiles. It is deemed that this new direction may offer a unique path forward in the direction of cost-effective lightweight composites.

This, indeed, is a very good time to be in composites!

REFERENCES

1. Cobbledick, D.S., Mong, F.M., Schreiner, B.A., Spencer, R.L. - Patent EP0461320- In mold coating compositions.
2. Schilkey, D.P., Hamner, J.R. - Patent US4867924- Method and apparatus for compression molding under vacuum.
3. Guha, P.K., Shah, M.S., Hamner, K.R. - Patent US 20150360425- Vacuum molding of thermoset sheet articles.
4. Guha, P.K., Siwajek, M.J., Yen, R.C. - Patent US20030100651- Reinforced polyester resins having increased toughness and crack resistance.
5. Guha, P.K., Siwajek, M.J., Hiltunen, M.J., Asuncion, M.Z. - Patent US 20150376350- Low-density molding compound containing surface derivatized microspheres.
6. Iseler, K.A., Wilkinson, R.E. - Patent US5130071- Vacuum compression molding method using preheated charge.
7. Alexandre, G., José, F. - Patent EP2349667- Device for converting materials using induction heating that enables preheating of the device.
8. Guha, P.K., Siwajek, M.J., Bonte, P., Toitgans, M.-P., Boyer, D. - Patent US 20200122413- Resin transfer molding with rapid cycle time.
9. Epel, J.N., Mcdougall, M.K., Wilkinson, R.E. - Patent CA1115299- Energy attenuator and method of manufacturing thereof.
10. Epel, J.N., Morse, III J.J., Trebilcock, T.N. - Patent US4707317- Method of making leaf spring.
11. Guha, P.K. - Patent US 20200139584- Process of making a fiber preform of commingled fiber bundle for overmolding.

7 Applications and Product Design
Composites vs. Multi-Material

Gajendra Pandey and Vamshi Gudapati

CONTENTS

APPLICATION AND EVOLUTION OF COMPOSITES BASED ON STRENGTH AND MODULUS

During the Stone Age, materials of choice were ceramics, glasses, and natural composites. With the industrial revolution, steel and alloys from cast iron technology were readily available and played a dominant role in design. However, the rate of development of new alloys slowed down in the 1960s paving way for innovation of new classes of materials in polymers, ceramics, and composites. Advances in polymer resins and glass fibers in the 1930s, along with the development of carbon fiber in the 1960s, laid the groundwork for materials used in fiber-reinforced polymer (FRP) composites.

Application of FRP composites first started and grew with boats. During World War II, military aircraft materials such as wood were substituted by lightweight, strong, weather, and corrosion-resistant composites. As corrosion of metals became an issue, corrosion-resistant composites were used in pipes and later adopted by the oil & gas industry. Development of composite manufacturing techniques such as pultrusion, vacuum bag molding, and filament winding enhanced unique properties of composites for high-performance engineering applications including large-scale rocket motors for exploration of space. With the proliferation of polymer composites, fiber-reinforced ceramic composites have become attractive alternatives to refractory metals and alloys. The use of ceramic composites has become popular for very high temperatures applications such as energy-efficient heat engines, high-subsonic supersonic airplanes, and spacecraft. The use of polymer and ceramic fiber-reinforced composites takes

DOI: 10.1201/9781003161738-7

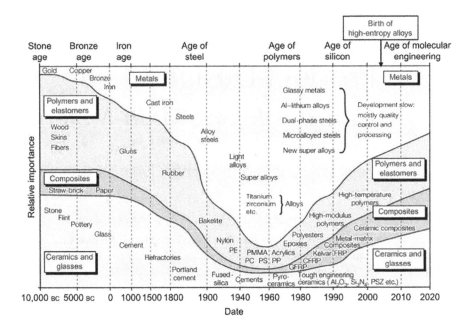

Materials selection in mechanical design, Third Edition By M. F. Ashby (Publisher: Elsevier Science & Technology).

advantage of high strength and stiffness to weight ratios combined with the flexibility in tailoring the matrix structure to meet the loading conditions resulting in greatly increased structural efficiency from low to very high temperatures. Advantage of composites materials in specific strength (i.e., strength divided by density), specific stiffness (i.e., modulus divided by density), and reliability under extreme conditions are helping them grow into new application market areas in defense, aerospace, automotive, infrastructure, and energy-related applications. With advent of nanotechnology, availability of new material choices, and rise of 3D printing, the need for stronger, lighter, and stiffer composites is expected to increase.

Usage of different materials in Boeing 787 [1].

Commonly used structural materials for commercial aerospace are metal and metal alloys, generally accounting for more than 90% of the weight of airframes. Aluminum alloy was a dominant material in airframe fabrication due to advantages in strength stiffness and lower weight. While metals and metal alloys still play an important role, the proportion of high-performance composites used in aerospace structures increased from less than 1% by weight in the 1920s to about 50% by weight in the new Boeing 787 and Airbus A350 aircraft. This trend was primarily driven by the ability of composites to decrease weight and increase fuel performance. For example, Boeing 787 uses 50% of composite materials in airframe and other primary structures and that offers about 20% weight savings compared to conventional aluminum designs (picture above). Weight savings translate into savings of thousands of dollars in fuel costs over lifetime for aircraft.

Lightweighting in the aerospace industry was primarily driven by global warming to reduce emissions and to increase energy efficiency. In addition to reduced weight and improvements in fuel efficiency, composites provide additional benefits in flight performance, high structural strength and stiffness, and improved safety performance. Like the aerospace industry, the automotive industry faces increased global competition for high-performance vehicles that reduce weight and costs combined with stringent environmental and safety conditions. A 1 lb weight reduction translates to $10,000 added value in the space industry, $100 in the commercial aerospace industry, and $5–$10 in the automobile industry. According to the U.S. Department of Energy [2], composite materials could have great potential to reduce the weight of a passenger car by 50% and improve its fuel efficiency by about 35% without compromising performance or safety. Vehicle weight and engine power are critical

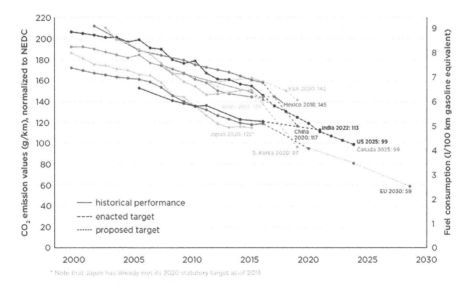

Zifei Yang and Anup Bandivadekar, "2017 Global Update: Light-Duty Vehicle Greenhouse Gas and Fuel Economy Standards" (International Council on Clean Transportation, June 23, 2017),

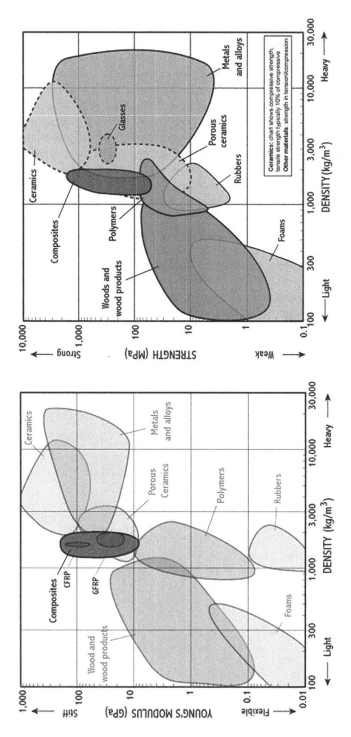

parameters that influence the fuel consumption and carbon dioxide (CO_2) emissions in the automobile industry. The European Union (EU) aims to reduce greenhouse emissions from automotive 40% below 1990 levels by 2030. Due to emission regulations in place for passenger cars in the EU, gas emissions reduced from 200 g/kg in the 2000s to 100–150 g/kg in 2020. New legislation in the EU is targeting emissions of 59 g/kg by 2030 with other regions expected to follow this trend. The proportion of lightweight composites used in automotive increased from less than 5% by weight in the 1970s to about 20% by weight in the 2020s. With the emergence of electric vehicles and more stringent regulations, the need for increased lightweighting will accelerate adoption of composites.

ADVANTAGES OF COMPOSITES

When combinations of properties are useful such as strength per unit weight and/or stiffness per unit weight, polymers and ceramics are as good as metals. Composites combine the attractive properties of other classes of materials while avoiding some of their drawbacks. Combining stiff, strong, and tough fibers with polymers and ceramics makes them significantly stiffer, tougher, and more damage tolerant. In both the aerospace and automotive industry, there is an increasing demand for tough, strong, stiff, and lightweight materials to replace metals and their alloys as structural materials. Notice from the graphs above, metals provide strength and stiffness at the cost of weight. Composites can be up to 70% lighter than metals and three to five times stronger. They have the flexibility to be designed as stiff or flexible and are well suited for applications that need high tension and increased service life with low maintenance. Unlike metals, they can be customized to provide strength or flexibility in critical areas while keeping the weight down where strength is not needed. In addition to structural benefits, composites also exhibit higher damping capacity and the ability to suppress vibration. Composites also exhibit nonstructural properties as they are nonconductive, noncorrosive, provide high UV resistance, have low CTE, provide EMI shielding properties, and do not rot. From a manufacturing and design process, composites are attractive when large complex shapes need to be molded replacing metal counterparts. Monolithic design of composites is desirable

	Composites	Steel	Aluminum	Plastics	Advantage
Density (gm/cm³)	1.2–2	7.75–8.05	2.7	~1	Up to 70% lighter than metals and allows precise weight distribution
Strength (MPa)	50–2,000	400–1,000	90–690	15–100	Up to three times stronger than aluminum and flexible
Modulus (GPa)	10–250	190–200	69–70	2–4	As stiff as metals
Part consolidation	Yes	No	No	Yes	Reduces maintenance and number of parts

over metal assemblies that are joined using traditional bolts or riveted joints. The number of parts that can be joined in automotive can be as high as thousands to a million in the aerospace industry. Monolithic composite design decreases assembly time needed for fasteners and joints in addition to providing reduced maintenance.

DIFFERENT CLASSES OF COMPOSITES

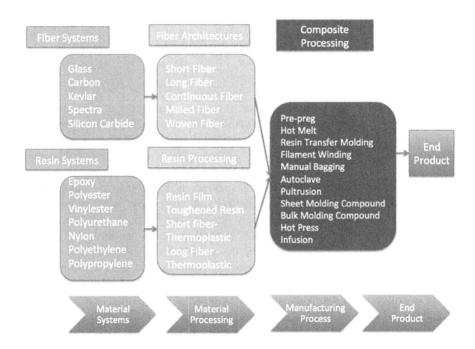

Fiber-reinforced composite materials are versatile and can be adapted for various applications based on end product requirements. Material requirements, manufacturing processes, and cost dictate the choice of composite used for a specific application. Continuous fiber composites are used for applications that are driven by high stiffness, modulus, and fatigue resistance. The fiber architecture can be unidirectional and or woven based on structural property requirements from the application such as wind blades, fuselages, filaments wound pipes, etc. The fiber-reinforced composite materials with the right combination of fiber and resin can yield the highest possible strength and stiffness at the lowest weight. By the addition of a lightweight core (honeycomb, foam, balsa wood, etc.), a very high bending strength and bending stiffness can be achieved without significant weight increase. Based on the fiber aspect ratio (length/diameter), a fiber-reinforced composite can be classified as continuous-fiber (large aspect ratio) and discontinuous-fiber (small aspect ratio) composite. The continuous-fiber composites usually have a preferred orientation (unidirectional, woven, etc.), whereas in a discontinuous-fiber composite, the fibers are randomly oriented (chopped fibers, random mat, etc.). In a continuous-fiber composite, the fiber volume can be as high as 70%. With further addition of fiber, there is not enough availability of matrix to hold the fiber together. Hence, fiber volume of higher than 70% is not recommended. In general, the applications where cost is not much of a concern then continuous carbon fiber is used for high specific strength and specific stiffness parts such as in aerospace application. Whereas, fiberglass composites are used in less demanding applications where one has to balance the cost and weight of a part. The composites are finding increasing applications in auto industries as well to meet performance requirements while reducing the weight for increased fuel efficiency. In the commercial transportation, cost is the major factor as composites offer lower weight and lower maintenance costs. Typically, fiberglass/polyurethane made by liquid or compression molding and fiberglass/polyester made by compression molding are being used. Due to durability and weight savings, the glass fiber composites have been used over traditional metals in recreational vehicles since long. Typically used products are fiberglass sheet molding compounds made by compression molding. In the automotive applications such as high-performance Formula 1 racing cars, where the cost is secondary the most of the chassis, including the monocoque, suspension, wings, and engine cover, are made from carbon fiber composites.

The gaining popularity of fiberglass in automotive industries for mid- to large-volume applications is due to its lower cost, lightweight, mechanical strength, corrosion resistance, thermal stability, dimensional stability, chemical resistance, moisture resistance, and abrasion resistance. Currently, in automotive, the composites are primarily used in front and rear bumpers, hoods, doors, and casings. They are also used in timing belts and V-belts, where glass strings are impregnated with rubber as reinforcement. The abrasion resistance properties are used in brake pads and clutches. Clutch disks are reinforced with woven fiberglass to maintain the integrity of the composite material. They are also used in both radial and bias-ply vehicle tire reinforcement; the glass cords are impregnated with up to 15%–30% resorcinol-formaldehyde-latex resin which coats the individual glass filaments. Another growing application of composites is in suspension components and drive shafts. For exteriors, the ultra-lightweight SMC continues to push sheet metal away. In some of the newer electric cars, the body is made using carbon fiber-reinforced plastics (very durable and light but expensive compared with fiberglass-reinforced plastic) that resulted in the body being 50% lighter than steel and 30% lighter than aluminum.

COMPOSITE PARTS DESIGN

The advantage of fiber-reinforced composite materials lies in their high resistance and stiffness compared to their low specific weight. This results in high specific properties compared to traditional materials like steel and aluminum. The composite materials also have excellent energy-absorbing capability per mass. Compared to single-layered steel in cars, multiple-layer composite laminates can be designed to absorb more energy in a crash. Modeling and simulation tools help designers optimize the design by predicting structural behavior and shorten the product development life cycle by reducing the testing and prototype development for the parts.

The composite materials are modeled primarily using (1) micromechanics-based approach, (2) an equivalent homogeneous material (EHM)-based approach, and (3) a combination of the two previous approaches. In the micromechanics-based approach, the material behavior is described locally hence, local defects, such as fiber–matrix detachment, etc. can be studied in fiber-reinforced composites. However, the computational time required to model such problems is very high because the mesh used is very fine compared to the EHM model. The EHM approach reduces simulation time, but it is not able to predict local effects, e.g., damage at fiber–matrix interface. A composite material's behavior can be separated into the microscale (interactions from constituent materials are examined in detail) and the macroscale (composite material behavior considered to be homogeneous and mean apparent properties are considered).

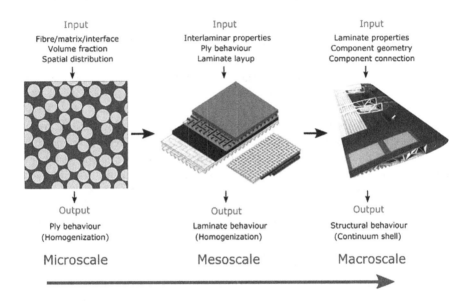

The length scale representation from microscale to macroscale [3].

The picture above shows the evolution of modeling from unit cells (microscale) that models the ply behavior from individual fiber, matrix, and the interface properties. The homogenized ply properties can then be used to build homogenized laminate behavior in mesoscale modeling. The homogenized laminate properties in turn used in the part modeling (macroscale) of a composite structure. In order to understand the failure mechanics, sometimes the microscale modeling is brought one level down to nanoscale modeling where the interface and the nano-additives are studied. This is called the multiscale modeling approach that is used to study the effect of constituents and microstructures on mechanical performances of the structure. In any part that has many components and constituents, it is impossible to model all the components in detail due to computational limitation. The multiscale modeling approach comes in handy in this case where the component of interest can be modeled at constituent level. However, this approach requires experience and careful judgment from the designer. Depending on the level of interest, the scale can be divided into micro-, meso-, and macroscale. Through this approach, the effect of fiber, resin, and the interface properties can be translated into the macroscale part mechanical performance. In finite element analysis, mechanical behaviors of heterogeneous materials are often described by using representative volume elements (RVEs). The RVEs are considered using either Hill's theory (1) or Drugan and Will's theory (2). The Hills's theory is currently being used widely where RVE is considered large enough to contain a large number of fibers and be a statistical representation of the composite materials. The effective material properties of RVE represent the material properties at the macroscale. This approach is commonly used for continuous fiber composites. For the discontinuous randomly distributed fiber composites, this approach of generating RVE is not commonly used because high-volume fractions and large fiber aspect ratios make it more difficult to model these micro-architectures [4,5]. Hence, a better approach to build RVEs for a randomly oriented discontinuous fiber composite is with high fiber volume fractions and large fiber proportions. The two most common approaches to generate RVEs for randomly oriented discontinuous fiber composites are the random sequential adsorption algorithm and the Monte Carlo procedure [4–7]. Another approach is the use of the automatic searching and coupling technique, where a 3D RVE is generated to analyze the composite with random fibers having a wide range fiber aspect ratio [7–9]. Currently, there are multiple failure criteria available for composite materials, namely Hill, Tsai-Hill, Tsai-Wu, Hashin-Rotem, Hashin, maximum stress, Hoffman, maximum strain, Hou, Puck-Sacihnürmann, Chang-Chang, Linde, LaRC03, LaRC04, Maimí, Hart-Smith, Yeh-Stratton, and others. From these the most commonly used failure criteria are the maximum stress, Tsai-Hill, Tsai-Wu, Hashin, and Puck-Sacihnürmann.

The conventional finite element software was developed for modeling isotropic material like metals. This leads to various challenges to model composite materials using the current finite element formulations and tools. Some of the challenges are:

- It is challenging to keep nonhomogenized properties of individual lamina for simulating larger parts
- The need for a very fine mesh to model individual fibers to capture the performance at the microscale. This is a very computationally intensive approach that is not feasible to use at macroscale.
- Convergence issues with the solutions. For some problems, the solver becomes sensitive to the convergence and requires a good level of understanding of the methods to overcome it.

Owing to the huge market demand of using these software packages for simulation of composite materials, the software providers started providing customized modules to solve such problems. For example, composites modeler for Abaqus, Ansys Composite Prep/Post, NISAII/COMPOSITE from Cranes Software, Inc., FiberSIM from Siemens, Helius-Composite from Autodesk, and GENOA from Altair and Laminate Tools from Anaglyph. The major limitation of all the FEM tools, as discussed above, is the output or results are as good as input or material properties. This is the major drawback of composite materials. Since, the composites are so versatile and provide so much flexibility to tailor the material for the desired mechanical properties that it comes with drawbacks. Even a small process variation could lead to completely different mechanical properties and failure modes. Another issue is scalability. For example, coupon-based tests are needed to evaluate the mechanical properties of a composite. However, scaling a coupon to a part level could introduce anomalies such as higher void percentage, improper fiber–matrix adhesion, improper fiber orientation, operator negligence, etc., that could result in the difference in the part performance compared to as designed using simulation tools. In-depth testing with material model development that can predict the part performance with a higher confidence level is required. One quick way to accommodate these differences between coupon to part level performance is to use higher values of safety factor that designers often use to knock down the mechanical properties. The safety factor as high as 35% or more is seen to be used. Higher safety factors result in overdesigning composite parts and that sometimes makes composites less attractive compared to metals. To overcome this, overdesigning two areas requires attention: (1) build a material model at the constitutive level to capture the properties of fiber, matrix, their interface, and all the damage mechanics that can be incorporated at the part level with minimal computational overhead and (2) the part processing needs to be precise enough that the coupon level properties can be reproduced at the part level and can be produced consistently without any deviation between the parts.

REFERENCES

1. http://compositeslab.com/composites-compared/composites-vs-aluminum/.
2. https://www.energy.gov/eere/amo/institute-advanced-composites-manufacturing-innovation.
3. Tan, W., Naya, F., Yang, L., Chang, T., Falzon, B.G., Zhan, L., Molina-Aldareguía, J.M., González, C., Llorca, J. The role of interfacial properties on the intralaminar and interlaminar damage behaviour of unidirectional composite laminates: Experimental characterization and multiscale modelling. *Compos. Part B Eng.* 2018, 138, 206–221.
4. Li, S., Sitnikova, E. *Representative Volume Elements and Unit Cells*, 1st ed.; Elsevier: Amsterdam, 2020.
5. Feito, N., Díaz-Álvarez, J., López-Puente, J., Miguelez, M.H. Experimental and numerical analysis of step drill bit performance when drilling woven CFRPs. *Compos. Struct.* 2018, 184, 1147–1155.
6. Tian, W., Qi, L., Zhou, J., Liang, J., Ma, Y. Representative volume element for composites reinforced by spatially randomly distributed discontinuous fibers and its applications. *Compos. Struct.* 2015, 131, 366–373.
7. Kanit, T., Forest, S., Galliet, I., Mounoury, V., Jeulin, D. Determination of the size of the representative volume element for random composites: Statistical and numerical approach. *Int. J. Solids Struct.* 2003, 40, 3647–3679.
8. Lu, Z., Yuan, Z., Liu, Q. 3D numerical simulation for the elastic properties of random fiber composites with a wide range of fiber aspect ratios. *Comput. Mater. Sci.* 2014, 90, 123–129.
9. Mechcomp3 *3rd International Conference of Mechanics of Composite*; Ferreira, A.J.M.; Viola, E.; Tornabene, F.; Fantuzzi, N. (Eds.) *Structural and Computational Mechanics Book Series*; Società Editrice Esculapio: Bologna, Italy, 2017.

8 Industry–University–Research Institution Partnership
A Case Study

Brian Pillay, Haibin Ning, and Kipp Carlisle

CONTENTS

INTRODUCTION

Composite materials have seen tremendous growth in global markets in all sectors of industry, automotive, aerospace, construction, marine, oil and gas, to name a few. This is well covered and documented by earlier chapters. In this highly competitive environment, the composite material community, business, academia, research labs, etc. need to look for a better framework that leads better understanding and furthers the use of composite materials, exploiting their many advantages over traditional materials in all of the above industry sectors.

Most universities do not have effective frameworks for engagement with industry and the reverse is also true, in that industry views universities as bureaucratic

DOI: 10.1201/9781003161738-8

institutions steeped in archaic policies that do not meet the challenges of today's business needs. Hence, there has been increased pressure on politicians to decrease state funding to universities, which has happened over the last few decades. State funding to most universities has reduced significantly with respect to the overall budgets, resulting in higher tuition and other fees and making university education inaccessible to many. The common interactions that businesses are used to with respect to universities are fundraising and donor opportunities, which rely on the generosity of companies/owners and their community engagement charter.

However, the authors believe there are opportunities for more productive and mutually beneficial collaboration between universities and industry, providing opportunities for income generation and more effective student education. Industry can benefit from utilizing cutting-edge research and expertise together with manpower development of next-generation engineers, which will make them more competitive on the global scale. This chapter will focus on the collaboration between the Materials Processing and Applications Development (MPAD) Center at the University of Alabama at Birmingham (UAB) and industry collaborators. The MPAD Center is a large (approximately 20,000 ft^2) material solutions laboratory with processing capabilities in advanced plastics and composites and metals casting. The center has industry scale equipment in processing and specializes in materials development all the way to prototype manufacturing. The experience gained from a project – development of a composite stop collar for offshore drilling application – will be used to demonstrate the opportunities and challenges faced. This project was a collaboration between Trelleborg Offshore (TO) and the MPAD Center at UAB. A nonmetallic collar was required to hold buoyancy modules on drill strings during operation in the harsh operating environment of offshore drilling. The collars were required on both ends on each string to prevent the buoyancy modules from floating off the drill string during deployment, use, and extraction of the drill string. The project involved materials selection, component design and analysis, prototype manufacture, and testing including field deployment.

PERCEPTIONS

The first key to establishing a working multifunctional partnership is the perception of the other members within each collaborating entity. The industrial, university, and governmental (if any) collaborators must all expect useful returns on their efforts, where each member's definition of useful returns can vary significantly. Setting and balancing these initial expectations will be the key to progressing the collaboration.

INDUSTRY

Industrial partners, when entering collaborative research agreements, are typically trying to satisfy pressing business needs. Perhaps the collaboration will be simply to solve a processing problem for an existing product, or to assist in developing a new way to process a recently commercialized material, or even to develop a recently commercial material into a viable product. In almost all cases, the industrial goals will be commercial, not altruistic, with the associated deadlines, budgets, deliverables, confidentiality, payback calculations, and liabilities. The main concerns of collaborating with academia arise where conflicts with the goals of academia occur.

Publication may often be discouraged due to confidentiality, especially if the intellectual property strategy of the business is handled via trade secrets instead of patents. These two concerns must be carefully balanced within the business to allow for collaboration with academia. One of the most potent tools to facilitate such is the ability of the business to monetize the publication via *marketing*. In a technology-driven business like composites, the importance of being recognized as the leader in a field, as evidenced by patents, publications, sales, and product track records, will have a value to the company and can be used to drive collaboration. In the modern environment of trade show exhibits and social media outreach, advertising the accomplishments of the company's collaborative ability and competence at *investing* in successful innovation is in itself a valuable outcome of a partnership that scientists and engineers should not overlook. Involvement and oversight of publications by the marketing team *in addition to* the company's technical process can assure that key advantage-providing details from a collaboration can be protected from competitor use, either through patenting or redaction from public disclosure.

The two other primary points of reluctance for the industry to agree to partner with academia revolve around cost and time. Industry frequently has development cycles of only weeks for minor improvements, to months for major ones, and true step-change innovation cycles that may encompass a few years. Unfortunately for academia, where students drive the tools and pace, even the largest corporations will devote only a small portion of their research and development expenditure to the long term. How, then, can effective partnerships be launched? Careful control of cost and delivery timeframes in the initial setup of the partnership may be the most effective. If smaller projects can be selected and structured to be closely related to other ongoing work at the university, multiple goals may be achieved simultaneously as the needs of several collaborators can be combined into the graduate education research of a single student. Key for industry will be establishing that smaller, faster timeline projects that can be achieved in universities by picking the right partners with ongoing research in areas of overlap and through more direct involvement by university staff.

ACADEMIA

Universities need to fully understand and appreciate the needs of their industry collaborators, i.e., goals, timelines, and budget, for the relationship to be mutually productive. Most research programs are geared toward long-term federally funded projects that span multiple years and no real pressure for quick data and analysis. These programs are geared more toward graduate student research and education with the view to publishing papers in groundbreaking fundamental research.

The huge stumbling block for effective industry collaboration is the understanding of university costs, bureaucracy, and timelines. Most universities view any project, no matter the size and scope, to be a full-fledged research project that should cover full facilities and administration charges with negotiated intellectual property (IP) agreements, etc. in place. However, the reality is that most times projects may be on a much smaller scale, with no potential IP, etc. The need for quick results/data and faculty expertise to help solve smaller problems in a timeous manner is typically more common. Universities should adjust their bureaucracy and expectations accordingly to have more meaningful and mutually rewarding engagement with industry.

Faculty members and their students should understand that smaller businesses do not care about the scientific merit of conducting long-term experimentation and providing complex solutions to the problems of the day. If solutions are not quick and effective, it means loss of revenue, which could result in bankruptcy of the business. Industry collaborators have often been overheard saying "we appreciate the academic credentials of Professor X, but we may not be in business any longer, by the time we receive the report of the investigation."

There definitely is a role for academic institutions to support industry in providing resources and capabilities to help solve problems in an effective and affordable manner. Universities need to understand that not everything that is done results in IP and patents. Most times building valuable and mutually rewarding relationships results in more income for universities rather than expecting to license IP. Most industries would like to support universities, especially in providing opportunities and training for students as part of their mission. Industries need to also understand that having a robust university partnership helps them develop and influence their future workforce. The majority of graduate and undergraduate students who have worked on industry projects ended up being offered employment by that particular company. In many instances, there was a competition by many companies for the student simply because the students (especially graduate students) were well versed with industry expectations, timelines, working on multiple projects simultaneously, and the demand of particle workable solutions for daily short-term problems. It must be stated that the authors do recognize and appreciate the need for continued academic research with long-term timelines, solutions, and goals; it is our belief that they can be done concurrently with a more effective and valuable experience for both students and faculty. University administration needs to recognize the importance of educating students for the betterment of the entire community, including industry, and relax archaic thinking and reform policies and procedures that promote better and more effective engagement.

The MPAD Center has active nondisclosure agreements (NDA) with over 50 different industry collaborators and provides services ranging from basic testing to large-scale research collaborations. The relationship with industry generally starts from requesting basic services like testing and characterization. The collaboration with TO was built over a period of approximately 15 years, starting off with basic testing, developing to more complex processing trials, and eventually leading to the clamp project.

RELATIONSHIP BUILDING

Successful collaboration as discussed in the last section will not occur without committed effort from individuals driving the company and university toward agreement.

Often this will be through individual interactions of combinations of alumni, former research advisors, or past colleagues. With this driver, an arrangement at the university side that promotes collaboration is the concept of fee-for-service agreements. Facility for a university lab to issue and accept short-term work almost as would a for-profit laboratory services company can establish the initial relationship at the business level in addition to the professional. In the case study to be detailed in Section Case Study – Landing String Stop Collar, several years of collaboration through simple fee for service contracts, which industry can treat exactly as a lab services fee, preceded the larger research project eventually initiated. Multiple times, TO contracted for 2–5 days use of laboratory equipment and personnel to characterize materials or process small amounts of raw materials, using techniques and equipment not readily available in industry. Ease of use, direct utility of data and materials purchased, and ability of the university to cope with a standard business purchasing/procurement process for these smaller collaborations built trust within the business that is working with the university was fruitful. Further, delivery was typically *faster and more complete* than when a single third-party commercial lab was used to provide a similar service, due to the varied backgrounds and expertise of the students, staff, and faculty.

For example, consider scanning electron microscopy (SEM) imaging of a sample using different imaging detectors. Historically, such work has been sent to third-party commercial labs, where technicians unfamiliar with the nuance of the materials use backscatter electron (BSE) imaging as part of root cause investigations, with the conclusion that their surfaces appear relatively smooth, as in Figure 8.1. When the same has been sent to a university, different conclusions are drawn, as all the detectors available, often more and of improved technology, are typically used to image the samples, as secondary electron (SE) imaging reveals finer surface details – Figure 8.2.

TM4000Plus 15kV 10.5mm x800 BSE M   50.0µm

FIGURE 8.1   SEM image showing less details on the surface of particles.

FIGURE 8.2   SEM image showing finer details on the surface of the particles.

TECHNOLOGY READINESS LEVELS

The technology readiness level (TRL), first proposed by NASA and subsequently used by the departments of defense and other agencies, is an 1–9 scale where 9 is the most mature technology. It is well known that most universities are very comfortable at TRL levels below 3, with industries' main focus in the 6–9 range. Universities are more geared to fundamental scientific research, which is one of the pillars of academia in furthering scientific breakthroughs. However, the technology demands of the 21st century (and beyond) engineer in industry have grown substantially over the past several decades, and it is the responsibility of universities to prepare scientists and engineers for the challenges of the industry in a globally competitive environment. There have also been major shifts in the industry with the constant financial pressure to continuously deliver larger returns to shareholders. This has led many companies, previously on the front end of technology development, to trim down resources in the research and development and focus more on the core business. This has been especially obvious during the great recession and other periods of poor economic performance. There exist opportunities for universities and industry to collaborate in the 4–6 range, to take research into industry and for industry to utilize the expertise of university faculty and capabilities to assist industry in conducting studies to solve more complex problems. This will result in mutually beneficial collaborations which will increase industry competitiveness and allow universities to better equip students with real-world industry training and education. The case study below demonstrates the collaborative effort in a university and industry partnership in materializing a concept (TRL 4–6) and advancing it to a mature technology (TRL 7–9).

Case Study—Landing String Stop Collar

The MPAD Center at UAB in partnership with TO was required to develop a clamping assembly solution for "landing string buoyancy" (LSB) – a concept that involved the attachment of buoyancy modules directly to a drill pipe in a particularly harsh environment. The purpose of the buoyancy was to reduce the overall weight of the "landing string," which allowed significant expansions in drilling capabilities with little or no modification to the older generation, existing equipment. This technology concept is patented (US PT 7,383,885) by the ownership of Landing String Solutions, LLC (LSS), which exclusively contracted with TO for implementation and manufacturing of the technology. Figure 8.3 shows a drill pipe joint with 12 buoyancy modules attached to it.

The buoyancy modules are secured to the drill pipe by way of two stop collars, which serve to prevent axial movement. The stop collar was a two-piece design using four threaded rods and associated nuts to provide the radial clamping force. The current clamp design is shown in Figure 8.4.

The stop collars must be able to withstand an axial force in the region of 6,000 lbs. The environment in which the collars must perform is olefin-based drilling mud with a density between 8 and 14 lb/ft^3, at pressures up to 6,000 psi and temperatures

FIGURE 8.3   The concept of buoyance modules around a drill pipe.

FIGURE 8.4   The design concept of clamps for holding the buoyancy modules.

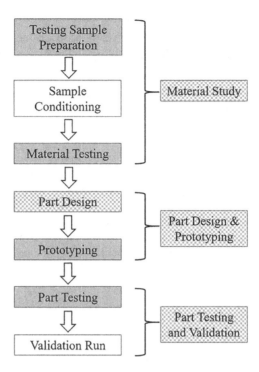

FIGURE 8.5   Flow chart showing the main tasks involved in the collaborated project.

up to 150°F. A further critical design requirement was that the stop collar must use no metallic components whatsoever. At the time, the LSB solution had the potential to revolutionize the drilling industry due to the tremendous cost savings and increased depth capabilities available to contractors and operators.

The flow chart in Figure 8.5 shows the main tasks in the collaborated project, from material study to part design and prototyping as well as part testing and validation. The tasks with the shaded background were predominantly done at the university with industry inputs. The tasks with the unshaded background were predominantly done in the industry with inputs from the university. The tasks with the checkered background were done at both university and industry with the same amount of involvement of both parties.

MATERIALS SELECTION AND TEST RESULTS

The operation condition under which the material works was very demanding for this deep-sea application. The hydrostatic pressure is extremely high due to the large depth that is up to 14,000 ft and can reach up to 41.9 MPa (414 atm). At the same time, the material has to stand an elevated temperature because the drilling mud is at the elevated temperature, approximately 150°F as mentioned in the previous section.

FIGURE 8.6 LFT glass/PA 66 pellets for molding the testing plates and stop collars. The pellets are 25.4 mm long.

Material study was first performed to select the material candidate best suited for this application. Several composite materials such as bulk molding compound (BMC) and long fiber-reinforced thermoplastic (LFT) composites with different matrices such as polypropylene and PA 66 were studied. The LFT composite consists of a thermoplastic polymer matrix and reinforcement fibers with a length above the critical fiber length (Figure 8.6). Because of the large fiber length in the LFT composite and discontinuous form of the fiber, it possesses several advantages, including ease of processability, excellent impact resistance, corrosion resistance, and good specific strength and modulus, which have made the LFT composite attractive for various applications such as automotive as a material alternative to its metallic counterparts. The use of the LFTs has led to weight saving and cost reduction without sacrificing the performance as a replacement of their metallic counterparts.

The materials were conditioned before mechanical testing was performed. The temperature and pressure used for the conditioning are listed in Table 8.1. Figure 8.7 shows the conditioning temperature and pressure with time for a single conditioning cycle. The samples underwent 14 such cycles. The material performance in impact, tension, and flexure was compared before and after the conditioning.

TABLE 8.1

Aging Conditions for the Testing Samples

Number of Cycles	Temperature (°F) Initial	Pressure (psig) Hold	Total Time per Cycle (h) Initial	Pressurization Rate (psi/min) Hold		
14	77	150	0	6,000	24	100

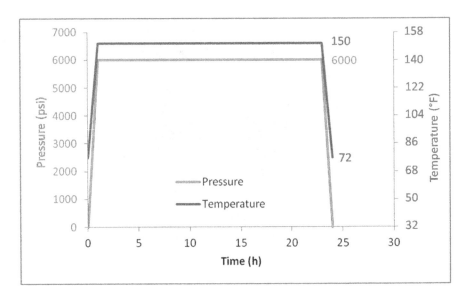

FIGURE 8.7 Graph of temperature and pressure for a single cycle of cyclic exposure.

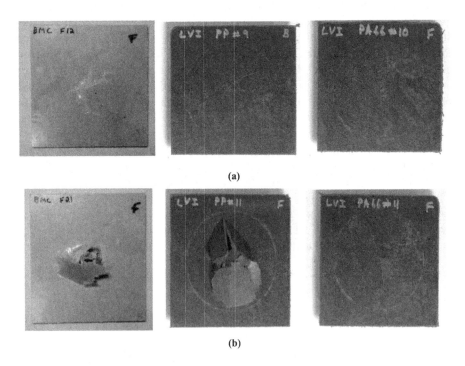

FIGURE 8.8 (a) Unconditioned samples tested at room temperature with 18 joule impact energy (from left to right: BMC, glass/PP, glass/PA 66). Note the larger damaged area for the BMC sample; (b) unconditioned samples tested at 60 joule impact energy (from left to right: BMC, glass/PP, glass/PA 66). Note the minimal damage on the glass/PA 66 sample.

Based on the testing results (Figures 8.9–8.11), glass/PA 66 showed the highest performance among all the material candidates before and after conditioning. Therefore, it was selected for the consequent prototyping and its material properties were used for modeling input.

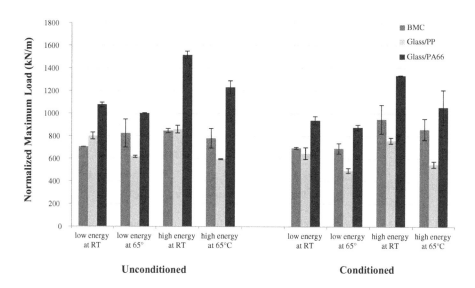

FIGURE 8.9 Comparison of normalized energy to max load for impact testing.

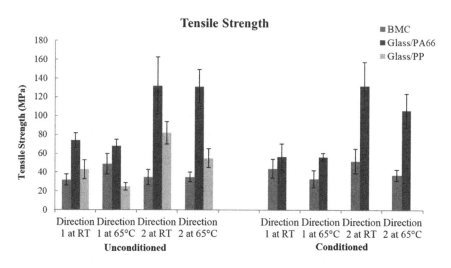

FIGURE 8.10 Tensile strength comparison for all of the samples before and after conditioning.

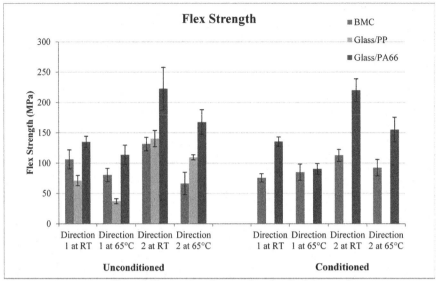

Figure 4: Flexure strength comparison for all of the samples.

FIGURE 8.11 Flexural strength comparison for all of the samples before and after conditioning.

Component Design and Prototyping

The stop collar was first designed using finite element analysis (FEA) to ensure that its performance meets the stress requirement in the application, even at the elevated temperature 150°F. Figure 8.12 shows a half model of the stop collar and drill pipe assembly consisting of two stop collars and one drill pipe section. The design iterations show the decrease of matrix stresses. The design in Figure 8.12d was determined to be the final geometry for prototyping.

Manufacturing Method and Component Test Results

Compression molding was used to retain the fiber length and achieve the maximum mechanical properties. A set of compression molds was designed and machined for prototyping the stop collar. Figure 8.13a shows the female mold and Figure 8.13b shows both the male and female molds mounted on a hydraulic press for compression molding the stop collar.

50 wt% long glass fiber-reinforced PA 66 pellets with 1 inch (25 mm) length were used (Figure 8.6). The pellets are composed of glass fibers and PA 66 matrix in which the fibers are preimpregnated. An extrusion–compression molding process was used to prototype the stop collar. The pellets were first fed into the extruder and the PA 66 matrix was melted. The screw inside the extruder rotated and created an extrudite with fibers well dispersed in the matrix for molding. A hydraulic press was

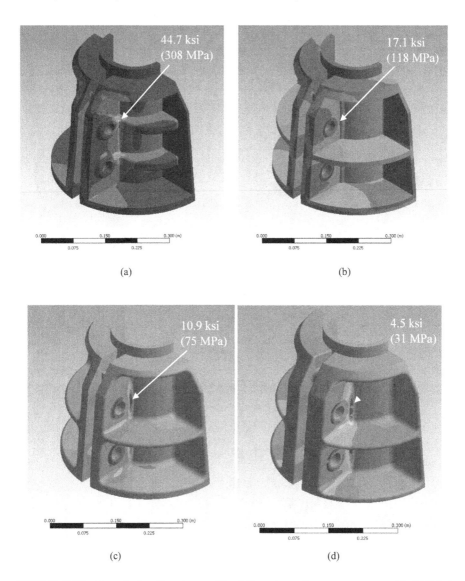

FIGURE 8.12 Part design iterations with different maximum stresses at (a) 44.7 ksi (308 MPa), (b) 17.1 ksi (118 MPa), (c) 10.9 ksi (75 MPa), and (d) 4.5 ksi (31 MPa). (Reprinted from Reference [1] with permission.)

used to apply compression force to the charge and consolidate it into the stop collar in a set of mold that is shown in Figure 8.13.

The molded stop collars were machined and then tested on a testing frame. Two stop collar halves were assembled onto a section of a steel pipe and the minimum load that initiated the slip was recorded (Figure 8.14).

The stop collar was developed, prototyped, tested, and met all of the requirements requested by TO.

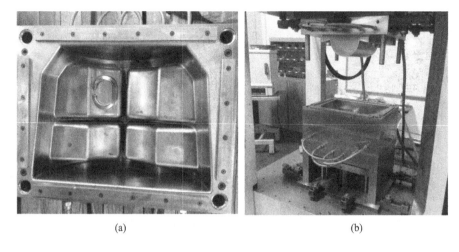

(a)   (b)

FIGURE 8.13   (a) The female mold for the stop collar and (b) the male and female molds mounted on a hydraulic press for compression molding. (Reprinted from Reference [1] with permission.)

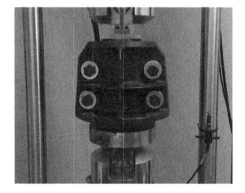

FIGURE 8.14   Slip test assembly with a set of stop collars assembled onto a section of a steel pipe. (Reprinted from Reference [1] with permission.)

HARDWARE

Commercially available composite hardware (nuts and bolts) were to be used by TO for this application. The hardware was manufactured using glass fiber and epoxy by a TO supplier. During testing conducted at the MPAD Center, the hardware passed all of the specified minimum requirements; however, the hardware did show signs of damage after repeated use. Personnel at the MPAD Center, who were working on thermoplastic composite hardware design, decided to develop thermoplastic hardware, as a research project, in parallel. The thermoplastic hardware were supplied to TO for parallel testing in hot–wet conditions and the performance was found to be significantly superior to the commercially available hardware.

(a)   (b)

FIGURE 8.15 (a) Buoyancy-enhanced landing string assembly being picked up for running in the hole and (b) a landing string being run in the hole with glass/PA 66 LFT stop collars being highlighted with red circles. (Adapted from Reference [1,2].)

LARGE-SCALE PROTOTYPING

All the relevant designs and test data were presented to TO's customer, LSS and their customers, and major oil and gas drilling companies. During the first commercial use of the buoyancy product, TO and MPAD Center at UAB jointly conducted a successful full in a field trial of the stop collar and hardware. This involved manufacturing 1,500 clamp halves and 750 sets of hardware. As a partner in the development with TO, the MPAD Center was again engaged in the process. As one could imagine, such a huge manufacturing task for a university laboratory was a cause for concern by both the company and university. The timeline of 1 year from purchase order to final delivery, with full quality buy off, was another huge challenge.

Because of the unique nature of the MPAD Center, the project was successfully completed, meeting all quality and manufacturing requirements, within budget and on time. MPAD Center personnel also assisted in educating and training TO and LSS company employees on correct installation methods and processes. The landing string, with the stop collars and buoyancy, was deployed and tested in the field as shown in Figure 8.15. The project was declared a success with potentially huge savings to drill operators in drilling costs and making the drill rigs safer.

CHALLENGES AND OPPORTUNITIES

As the reader would expect, the large-scale prototyping was met with huge challenges by both the MPAD Center at UAB and TO. The initial skepticism of company executives regarding engaging a university on such a large-scale multimillion dollar project with such tight timelines and industry standards and protocols had to be overcome through a combination of discussions, negotiation, and plan execution. The university management was equally concerned with the issues of liabilities with respect to product use and meeting quality specifications and timing. The attorneys from both the company and university went through several meetings, contract iterations, and negotiations. The reason the project survived all the scrutiny was due to the relationship built, over the years of collaboration, between the faculty of the MPAD Center and engineers at TO. Both parties had confidence and trust in each other with respect to expectations and support.

The opportunities presented to the students working on the project at the MPAD Center were so significant that every student involved in the project had multiple job offers. It should be noted that the entire project was conducted by two faculty members (who had other faculty duties and responsibilities), one full-time engineer, and ten graduate and undergraduate students working on a part-time basis. All materials inventory, manufacturing, quality and inspection, part inventory, and shipping were set up to supplier required guidelines. Needless to say, aside from the financial benefits to the students, the experience gained was unparalleled.

CONCLUSION

The case study shows that if universities and companies can step out of their respective comfort zones, successful collaboration can be achieved and it can offer limitless possibilities to both industry and universities. It is the experience of the authors that the fundamental requirement for successful collaborations is the ability to build relationships that promote trust and confidence in capabilities and expertise. Needless to say that honesty and integrity on the part of the individuals involved in promoting and fostering the collaboration is the cornerstone of the partnership. It is the opinion of the authors that a willingness to be entrepreneurial on the part of both universities and industry will yield mutually successful collaborations that not only benefit both entities and further their respective missions but also benefit students, who are the next generation of scientists and engineers who will be required to solve the world's biggest challenges.

ACKNOWLEDGMENTS

The authors wish to acknowledge the management of Trelleborg Offshore, Landing String Solutions, and The University of Alabama at Birmingham.

REFERENCES

1. Pillay, S., Ning, H., Barfknecht, P. and Carlisle, K., 2020. Long fiber thermoplastic composite for offshore drilling application - Design and prototyping. *Composites Part B: Engineering*, 200, p.108339.
2. Eberstein, W., Eberstein, C. and Pattarozzi, E., 2016. Buoyancy system lands nearly 20,000 ft of 14-in. casing in GOM. *World Oil*, 2016(July), pp.77–80.

9 The Customer…

Jim Plaunt

CONTENTS

This is personal; it was my life for 40 years.

I entered the composites industry as an hourly employee. To augment the meager salary of an academician, I worked at an automotive OEM composites plant during the summers. My primary task was to dump 50 pound bags of calcium carbonate into an SMC mixer.

Eventually, I left my tenured faculty position to join the composites industry. My academic friends thought I was crazy; "nobody leaves a tenured position!" Nevertheless, I found the production environment fascinating and I set about to learn as much as I could about my new world. At the OEM, I wore the following "hats": hourly supervisor, compounding manager, molding engineer, and engineering leader. Unexpectedly, my OEM employer exited the SMC business; I remained with them for the next 3 years, and soon after the launch of the Taurus, I was asked to join a material suppler that provided resins and reinforcements to the SMC industry. I accepted the position and entered through their technical service organization.

I want to highlight the word **service**. You must do your best to provide all your customers with **great service**.

Thoughts regarding great service will be interspersed throughout the remaining dialogue. But here are some of my personal mantras.

1. Be positive and polite. In the customers' eyes, you are most likely the "face" of your company. It should be part of your life; live it. Carrying a negative outlook is a recipe for disaster. Do all you can to turn a negative into a positive. In some cases, you will have to fight for your customer within the ranks of your own company. If that is the case, you will need to formulate a direct and comprehensive strategy to fight your internal battles. Stay strong and dedicated to your craft.
2. Go beyond the standard expectations. Exceed expectations. Have a plan. Exceed that plan. Within your own company and with your customers. Leave them with a sense that you really care about them. Great service will

DOI: 10.1201/9781003161738-9

make you someone special and you'll likely exceed expectations; your customers will look forward to seeing you!! Go the extra mile! However, this
comes with the caveat: if you're not willing to work at it, it's probably not
going to happen.

3. A promise made must be a promise kept. If you said to a customer that you
will do something, then you must do it. That is imperative. No matter how
big or how small; if you do not follow up, it will be remembered. Emails and
messages of resolution are important. Always document your visits to keep
a history of your interactions.

4. Customer complaints: For God's sake don't blow them off. Apologize to
your customer. Get all the information concerning the issue that you can.
If your contact is relaying the complaint, try to access the source of the
complaint. Understand the complaint. Write it down; document it. You are
responsible for solving this complaint. Get after it right now; take action.
Thank your customer contact for advising you of the issue.

5. In the configuration of many sales organizations, the individual sales personnel are often working from a remote "office," quite often their home.
This state-of-affairs leave the salespeople on islands and remote from their
respective organizations; "island living." Gone are the days of sales offices
where a sense of team/community was often present, where discussions of
concerns or times of celebration were held. So, to those persons who work
in labs/manufacturing facilities and other business locals staffed with multiple personnel, please be cognizant of those of your brethren who live on
their personal islands and attempt to include them in as many activities as is
possible.

6. In reference to "island living": I believe it's imperative that there is an active
dialogue between the sales organization and the technical community of
your company. Joint meetings should be held on a regular basis so the sales
personnel can share the wants/needs of their customers with the technical
organization.

AND NOW SOME INSIGHTS AND REMEMBRANCES

It's important to realize that you have a multitude of customers within your sphere
of influence; they reside both inside and outside the company of your employment. It
will be in your best interest to engage positively with each one of them and on as an
individual basis as possible.

It's of value to know all of them as well as you can; to understand what they enjoy;
what's important to them; what they like and what they don't. When you sit in their
office, their place of business, look around see their pictures, family, golf, and an
outdoor person.

My caveat to you….as you deal with your "customer" is to try and find a personal
base that is a positive, something that you can discuss outside of your working relationship; if so, things will likely move along on a smoother tact if you can establish
that relationship.

After many years in technical service, I moved into technical sales and….

I had a customer, a respected purchasing agent, who was afflicted by polio in her youth and only had the use of one arm. In our business discussions, to my amazement, I became aware that she was an avid golfer; she swung the club with one arm! She was adamant that she could overcome any physical issue she faced…..or any sales guy that crossed her path (me). We enjoyed our times on the golf course; there were not many of them, but the times we spent on the course were important. Also, over an occasional lunch, I learned that she was an aspiring artist, watercolors, and she anguished over the concern if she was good enough to display her work in juried art festivals. It was an interesting juxtaposition to see her "strength" as a purchasing manager and her self-doubt as an artist. I saw a representative piece of her work; she was very good!! I encouraged her to work on her obvious talent and she eventually enjoyed success in her artistic endeavors. A large watercolor painting of her exceptional talent hangs on a wall in our home; it is representative of her seemingly limitless capabilities. She was one of a kind, as are all of your customers. Over the years, our fearsome negotiations continued, but we had a positive relationship upon which we could always find balance. Negotiating in good faith is an aspirational goal.

But in our business, in the composites industry, our customers are as varied as are our products. We must deal with all levels of the manufacturing/management systems in which we are thrust.

Of course, at each specific customer, there is an inherent hierarchy and that structure is important to understand but to dismiss the "worker bee" to only laud the "big kahuna" can be detrimental to potential success. I have always found it important to win the heart of the "soldier"; it helps build a positive consensus when questions are asked. It's difficult to win if the person doing the (physical) work is not a positive respondent to you and your product.

The automotive arena in which I existed was bound by iron-clad specifications; hoops to jump through; regulations to understand. Without the knowledge of those standards, it's difficult to navigate a successful passage. So it is incumbent on you to understand the rules; you will likely be the individual to assist in leading your product development internally and to be on the forward edge of its evaluation at the customer.

Listen, listen, listen and write it down; take notes; don't go into any meeting without a pen and paper!!

And data, data, data; understand the importance of good data and where it is applicable. Be able to present it in fashion that your contact can understand it….. and….also, make sure you understand it.

In the world of composites, and specifically *automotive* composites, it's a multitiered world. As a supplier to a Tier One customer, you have to navigate the Tier One structure and have a firm understanding of the OEM's wants and needs. It is imperative that you have an in-depth understanding of the technology you represent and how your product meets the needs of the OEM. Often the Tier One supplier will not want you to have access to the OEM as they desire to control that interface. If it is possible to have a working relationship with the OEM without upsetting your Tier One supplier, it will be beneficial to your success. In all honesty, it will take an extensive amount of time to gain the trust of both parties; dogged determination and persistence must be part of your DNA.

As to securing new business at an automotive OEM, IT'S DAMN TOUGH. It's a marathon. You will be challenged many times throughout this endeavor; never fall under the belief that you have covered all the bases......because the phone will eventually ring and you will, once again, be under duress; it's not for the faint of heart, but it is invigorating.

Your company must understand that once your new composite material is approved by the automotive OEM that it could be 4 years before you will see it on an automobile in the marketplace.

Patience grasshopper patience.

However, once you are on a platform it'll be difficult for you to be replaced as now you are "spec'd in." And that long road to success will remain difficult as you and your company must continue to produce a consistent product; you will be constantly challenged to improve quality and reduce pricing.

ANOTHER REMEMBRANCE

My multilayered agreement was about to expire at my largest customer. The agreement involved technical activity, quality standards, delivery requirements, and pricing on a multitude of products. The purchasing manager and I had known each other for decades. He was a big bear of a man; he could be an imposing and intimidating figure. We started our negotiations far in advance of our termination date.

I heard all the threats:

"We'll certify another vendor!!"

"The OEM is demanding a price reduction!!"

"You are jeopardizing my position if I can't get your pricing down to a reasonable place...I'm really under the gun here!!"

"If you and I can't get this done then I'll have to get your Upper Management involved!!"

This went on week after week after week.

My responses:

"We've been an exceptional supplier."

"Our service and quality are unparalleled."

"Our technical teams work well together; we're working together on new Class A material improvements."

Week after week after week.

Charts. Details.

Discussion after discussion.

We finally came to an agreement. I was relieved to have successfully negotiated the new contract and advised my corporate office to draw up the terms of the agreement; sign it and get it to me in an overnight service and I would take it in for his signature. I had successfully negotiated a contract that was positive to my companies' desires.

Success is sweet.....

Within 24 hours, I had received the signed documents from my corporate office. I called my customer and told him I'd be in the next morning at 9AM to finish up this marathon. He responded affirmatively.

I was at his office at 9AM. Without a word and with a smile and a satisfied flourish, I placed the documents on his desk in front of him for his signature. He picked them up and looked at me and said: **"I can't sign these."**

I was stunned!!

Speechless…

I did not know what to do/say….nothing.

I grabbed the documents out of his hand. Grabbed the coat off the chair and stormed out of his office. In the parking lot, I threw the documents and coat in the back seat and headed out of the parking lot. I was beyond upset.

Now, I would have to call my headquarters and tell them that The Deal was off!! And that I'd have to get the details in a day or two. But at the moment I'm beyond upset…..and trying to figure out a strategy for my call.

I'm on the freeway; cell phone rings.

I answer it.

It's him.

He says: "Are you OK?"

It takes me a moment to respond.

"Yep, I'm OK just ticked off……."

He says: "Why don't you bring **my coat** back and we'll work this out."

I had inadvertently taken his leather coat when I stormed out of his office.

The issue with the new contract was a minor misunderstanding about Terms; which we ironed out quickly.

For many years, we laughed about me "stealing" his coat.

I must also profess that past technical/in-plant work experience can be of immense value. If you have walked the journey of the customers you service, you have the potential to gain immense credibility. However, you must be careful not to suffer from the "world's smartest men" syndrome and continually profess your brilliance; that gets old in a hurry.

What I did with my experience:

As you may have noticed in my Bio information, I spent a number of years in the academic arena; I enjoyed teaching/coaching; it was to be my life work…..until I found composites.

A QUICK FLASHBACK

To pay for my university education, I worked on the Canadian National Railway (CNR) during every summer vacation and with some regularity during Christmas breaks. I worked in the signals department (crossing protection, communication maintenance, block signaling, etc.). Having a job on the railway was fortuitous as they paid very well and I was able to get through school with very little debt.

During my last summer on the railroad, we rewired (switches) a rail yard in Toronto. A couple of weeks into the job I was invited to a retirement party for

"Chuck," a longtime CNR employee. Free beer. As the celebration drew to a close the section supervisor pulled "Chuck" to the front of the room to say a few words and present him with his retirement gift. The retirement gift was an inscribed gold railway pocket watch, attached to a gold chain; such gifts were much coveted by railroad retirees. The supervisor called for quiet; the room hushed. With his right arm around "Chuck's" shoulders, the supervisor gave a few words of introduction. During his introduction, the supervisor was gently swinging the gold watch, on the chain, in front of "Chuck." Each time the watch would swing in front of "Chuck," he would attempt to reach for it; it became a bit of a game. As the supervisor was ending his remarks, he said to "Chuck": "I know you want this watch....but before you get it I'd like you to tell these guys a few things that you've done during your thirty years on the railroad." "Chuck" slowly went through his career and, in closing, said that'd spent the last year working in this yard. …..Pause….. The supervisor casually asked him what he specifically did during the last year.

With some pride, "Chuck" said that he personally checked every locomotive that left the yard during his shift. The supervisor asked him to elaborate on what he did to "check" every locomotive. "Chuck" explained that with his hammer he struck *every* wheel of *every* locomotive that left the yard (that's a lot of wheels). This time the supervisor paused and after a moment of silence said to "Chuck": "Chuck, that's great....now, why did you hit every wheel with your hammer?" Chuck shrugged his shoulders at said: **"I don't know."** The supervisor gave "Chuck" his watch and there was free beer.

I didn't drink so the enticement of free beer did not appeal to me but "Chuck's" response of **"I don't know"** bothered me greatly. I'm a young 20-year-old, about to graduate from college and I just heard someone say he was doing a job and he didn't know why he was doing it!!! I was stunned…….

Some years later as I was working on the production floor in an engineering capacity for an automotive OEM, it became very clear to me that most of our hourly employees simply did **what** they were told to do but often they did not know **why** they were doing it.

This eventually led to the development of an 8-hour seminar: **SMC 101- From What to Why** that I have taught multiple times in many of the corporate offices and manufacturing plants in the industry. Explaining the basics of formulations, best molding practices, and troubleshooting. I was always pleased that the management of these facilities would provide me the opportunity and time to instruct/train their workforce. From French-speaking molders in Quebec to young Chinese engineers to SMC operations in Minnesota/Wisconsin/Louisiana/Michigan/Ohio/Indiana/North Carolina/Texas, I have spread the gospel in hopes that by improving the basic understanding of their jobs they would produce a better product. Education is never wasted.

One more admonition: **be on time…….do not ever be late; never, never be late** for a meeting with any customer. My goal was always to be in the parking lot at least 30 minutes before any meeting. This would allow me to go over my notes and item list. Preparation is absolute. Preparation will allow you to be prepared for the surprises that often occur; it will provide you with a solid base to work from.

And if there is a question you cannot answer; it will be to your advantage to say: "I'll look into that and get back to you." Do not try to bluff your way through an unknown. Make sure you write down the question/issue/concern. And look after it…..look after it…..as soon as possible. Questions that go unanswered will become problematic.

IT DOESN'T ALWAYS WORK OUT….

At this point, the resin/reinforcement company for which I was employed spun off the resin portion of the business into a joint partnership with a privately held resin company located in Tennessee. I made the decision to cast my lot with the new joint venture.

The company that we "joined" was not involved in the SMC industry but mainly focused on marine, spray up, and other sundry resin businesses. They were well known for their aggressiveness in the markets they served; decisions were made quickly; current businesses were often serviced through long-term personal relationships.

In the consolidation of the two companies, along with us came a major SMC customer; a large bathware manufacturer located in Alabama. This business was "legacy" business and was serviced from/by our primary SMC resin producing plant located in Indiana. Technical service was also provided by the Indiana facility and sales coverage from the corporate office. Much of the formulation technology for the bathware company had been developed in concert with my "old" company.

In the transition, to our new joint venture, the sales responsibilities for the bathware company were transferred to the folk in Tennessee; it made logical sense as they were geographically closer to the customer; it was also thought that, eventually, the resin manufacturing would be transferred to a southern production facility operated by our new partners.

Soon the southern sales manager and sales representative introduced themselves to the plant manager of the bathware facility. This plant was one of the largest SMC facilities in the world, and as a result, their use of resin and reinforcements were representative of its size. They were loyal; they were long term; they were important. We had history on our side. From all indications, the meeting between the new sales team and the bathware plant manager went well and the responsibility to handle the account was transferred to our new partners in the south.

Coincidentally, at this same juncture our bathware customer was in the throes of announcing their first **Global Supplier Award of Excellence** in the history of their company; it was a big deal. And as luck would have it we found out that we were in the final three selected. We were elated!! Let me clarify: our "old" company was in the final three. But, nevertheless, our new joint venture was our representative in the final three.

Soon we came to understand that the meeting between the bathware plant manager and our "new" team did not go as well as we were led to believe. The plant manager felt that his facility (and he personally) was being disrespected. He quickly came to the understanding that our new team had absolutely no SMC experience. A hallmark of our past relationship was that his account was serviced by highly qualified individuals that had a long history of interaction with his facility and its processes. To put a "nail in the coffin," our new team *did not wear socks!*! to their meeting. He felt they were far too casual and "unqualified" to service his account. He was very upset.

He was further upset by the fact that now.....a supplier that he feels let him (and his plant) down....was being considered for the first **Global Supplier Award of Excellence** presented by his company. He seethed for a while and then called his old contacts within our new company and made his complaints known. Quickly a change was made in the sales coverage and I was appointed to handle the account.

I requested a visit as soon as possible.

Within 48 hours, I flew to Alabama and was sitting, uncomfortably, in the plant managers' office. We had a "come-to-Jesus" meeting. He remained displeased. So much so that he had contacted a competing resin supplier to potentially bid on our business. Yikes!!!

We had been the supplier of record to this facility since its inception and could not envision the possibility that we would be replaced at this account. And, to make matters worse, if we lost the business it would leave a gaping hole in the output of our manufacturing plant in Indiana. I had a lot on my plate.

Within the next couple of weeks, I spent an inordinate amount of time at the bathware facility. I met with the plant's technical manager. He was an exceptional engineer and had developed unique mixing equipment for their compounding facility. It was an "exotic" extrusion system (no mix tanks) that ran directly to the SMC machine.

An interesting note: the fiberglass reinforcements were shipped to this facility in four hundred (400 lb.) doffs!! A standard doff size is around forty pounds.

The technical manager was in the process of bringing a newer, more refined system (based on the current system) that could handle the increased demands of future business. So, my time with him was hindered by the necessities of the labor he had to expend on his new equipment. But, he was knowledgeable of the plant manager's dislike for my company; and he had been directed to work with our competition to replace our products. A technical representative of our competitor was already onsite!!

To my surprise, my company's leadership team was invited to Wisconsin for the gala presentation of the first **Global Supplier Excellence Award** program; I was included in the travel team!! We were in the final three. Obviously, word of the bathware plant manager's unhappiness with our firm had not reached their corporate headquarters.

We won!!!!

We were the inaugural winner of the **Global Supplier Award of Excellence**.

David, the customers' President and CEO, upon announcing our win to the audience expounded on our capabilities and of our importance to their company. Fred, our president, graciously replied and mentioned that we looked forward to many more years of successfully working together.

Meanwhile, back at the bathware molding plant:

A. The competition trialed a successful competitive product
B. We were leveraged for pricing

Ninety days later, we had no business with them.

In retrospect:

1. We took the customer for granted.
2. We did not provide them with the service they deserved.
    A. The new sales personnel who did not understand their business.
    B. They did not wear socks.
3. Do not underestimate the capabilities of a vengeful plant manager.
4. To my knowledge, the **Global Supplier Award of Excellence** was never again awarded.
5. We used our reactors to produce products that provided higher margins.

Thank you for taking the time to read through my thoughts and remembrances. It has been a personal honor to be included in this manuscript.

Wishing you all good health and continued success...............

10 Composites New Business Development

Navigating the Not-So-Obvious

Cedric Ball

CONTENTS

There are few references describing the process of how composite materials are selected for an application. Rarely, does a first commercial sale happen organically? Technical products, like most other products, do not sell themselves. They have to be sold no matter how well the new product solves a problem or addresses a need. Likewise, producers of composite materials have to make deliberate efforts to find new applications. This process is called by several names. Business development, new product development, business growth, innovation, and monetization are among their many references. Simply put, their common aim is to sell more products and to improve profitability for the enterprise. Business development, as called here, is the space between the marketing and sales activities in the commercialization process. It is where marketing departments develop value propositions based on the common needs of customers. Sales teams execute on selling the new products or services that their marketing teams develop. This process, the business development process, is the most critical function for a business' long-term growth. It is also the most difficult!

DOI: 10.1201/9781003161738-10

The composite materials' business development process can take anywhere from 1 to 10 years, or more, depending on the end-use market segment. The more highly engineered the product, the longer the development is likely to take due to various performance requirements and safety-related qualification testing. Business development, best enacted, is embedding oneself in the new business development process of the customer who is incorporating composite material into some aspect of their new product.

Industrial organizations typically use a Stage-Gate® type approach for developing new products [1] (see Figure 10.1). The composites business developer interacts most heavily during the concept development stage when the customer is considering which material to use for their application. When the situation is a first time metal-to-composite conversion, the customer must establish proof-of-concept of the component made with composite. The concept can advance toward commercialization only if the "proof" tests are successful. Proof tests are intended to challenge the most critical aspects of the composite material in the context of the application. The context can be special load cases, environmental conditions, or a durability test.

In addition to technical proof tests, end-users will also prepare financial estimates. The cost of a new product is as important as its performance aspects. These two characteristics are inseparable. New products must work; they must also be affordable. Therefore, a business case showing financial feasibility is part of the early Stage-Gate process. The business case can be as simple as a cost comparison between the composite and incumbent solutions (see Table 10.1). They require more detail as assumptions are verified and the new product progresses toward full-scale commercialization.

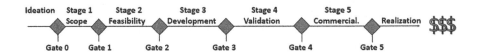

FIGURE 10.1   Typical new product development/industrial Stage-Gate® process.

TABLE 10.1

Simple Business Case Showing the Cost Savings of a Composite Part versus Incumbent Steel Design

	Example Design	Proposal	Savings/Benefit
Material	Steel	Composite Design	
Approximate part weight	1.4 lbs	0.9 lbs	35%
Estimated piece cost	$24.00	$13.00[a]	45%
Annual volume	125,000	125,000	
Annual piece cost	$3,000,000	$1,625,000	$1,375,00

[a]Assumes 2-cavity production tool and post-baking step. Does not include trim bezel.

CHALLENGES AT THE END-USER

Composite materials routinely face several business development challenges. These include:

- Engineers and decision-makers that have insufficient understanding of composites as compared to metals
- A value chain that is unfamiliar – perceived to be long and complex by end-users
- Purchasing and engineering organizations that are structured to optimize incumbent material solutions
- Decision-makers that have disparate and conflicting incentives

UNDERSTANDING OF COMPOSITE MATERIALS

Composites are perceived by many to be difficult to engineer, expensive, and often ill-suited for mass production. These perceptions are among the first obstacles that need to be overcome by the composites business developer. End-users must be convinced that composites are not only the best technical solution for their application. They must also be convinced that composites can be affordable. In most cases, composite raw materials are more expensive than metals on a pound-for-pound basis. However, it is the cost of the final part and not the cost of the raw material that is relevant. A composite part can be less expensive than its metal counterpart due to its lower density and more efficient material use. In too many cases, composites are dismissed as a viable alternative because they are not compared at the part or system level. Therefore, when sources report on the various obstacles to lightweight materials, the specific circumstance and experience of the respondents must be taken into account [2] (see Figure 10.2). Are the respondents familiar with composite design? Is the comparison being made on a part-to-part basis where the composite design has been optimized? Of the many composite materials and processes, which ones are being compared? Given that most engineers are not experienced with composite materials, overgeneralized assumptions can lead to wrong conclusions. The business developer must confront these common misperceptions.

The composites business developer can expect to spend months, even years, to educate engineers and key decision-makers. The more complex the composite substitution, the greater will be the need to educate customers about the details of composite material performance and its differences compared to metal. For example, there is the fact that most metals have isotropic thermomechanical behavior. Engineers can take design shortcuts when this is true; parts can be designed accordingly. Continuous fiber and short fiber-filled composites, on the other hand, are anisotropic and must be designed with specific load paths in mind. Directionality matters and process has a large influence on final part properties. A well-designed composite part puts material in the specific locations and directions of the loads acting on the part. All other material can be omitted resulting in a design that is lighter than its metal counterpart. These principles and others must be explained to engineers who are inexperienced with composite design and their manufacturing techniques. Live demonstrations are

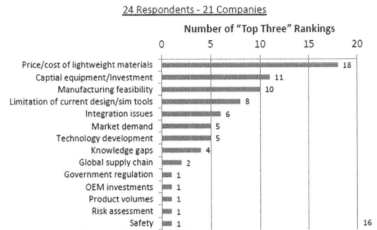

Barriers to Implementing Lightweighting Technologies
"Top Three" Ranked Responses
24 Respondents - 21 Companies

FIGURE 10.2   [Perceived] Barriers to implementing lightweight technologies.

often required to convince engineers about the feasibility of high volume manufacturing and part performance. Else, the composite part tends to be overdesigned. The inexperienced engineer uses safety factors that are too conservative leading either to parts that do not perform or that are more expensive than necessary.

Materials engineering textbooks routinely devote only a paragraph or two covering composite materials compared with several chapters for metals. Knowledge in thermosetting composites is especially lacking due to the added level of complexity concerning thermoset chemical dynamics. Therefore, career engineers who go on to have the decision-making authority to use composites in a new design may not be aware of their options. Lack of knowledge equates to higher risk for the engineer. In addition, since most can afford neither the time nor the resources to learn new materials on the job, they are unlikely to consider composite material as viable for the application. Engineers use carryover designs as much as possible unless challenged with new performance requirements such as lightweighting. Business developers are faced with overcoming the mantra, "Don't fix what ain't broke!"

However, the flipside of this challenge offers some opportunity for business developers. Although scarce for time, engineers are natural learners. They are curious and even competitive when it comes to gaining knowledge. Business developers can conduct training seminars, lunch-and-learn sessions, product demonstrations, and field trips. All these are ways to educate engineers in a low-pressure environment.

THE COMPOSITES VALUE CHAIN: PERCEIVED TO BE LONG AND COMPLEX

Concerning the composites value chain, it is a fact that the value chain for composites is long and complex when compared to metal commodities (see Figure 10.3).

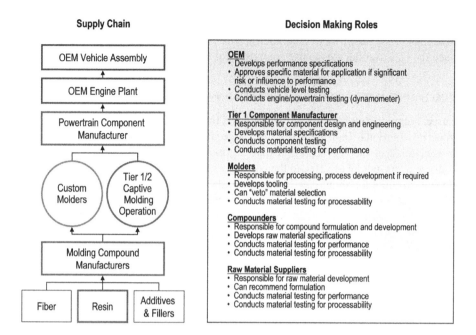

FIGURE 10.3 Composites value chain and decision-making roles.

Lower volume products tend to require longer and more costly supply chains but then streamline themselves as higher volumes justify consolidation and dedicated channels. End-users usually buy molded composite parts to assemble with other parts. The molder buys compounds, resins, and reinforcements from chemical and fiber producers. The chemical and fiber producers buy various raw materials to produce their respective commodities.

The end-user of a composite material is usually not the molder or fabricator of the composite part. However, the end-user needs to qualify its use. If the end-user decides that she is uncomfortable or restricted from using composite materials in the application, then the possibility for business along the entire value chain stops there. Understanding the complete value chain is important for establishing an effective business development strategy and a key reason to engage with the OEM end-user.

For example, a municipality that wants to install a lightweight composite bridge must approve the use of composite materials (on behalf of the bridge users). These approvals often come in the form of codes or material performance specifications that have undergone extensive testing for safe use. Traditional materials such as iron, steel, or concrete have demonstrated their performance over decades. Composite materials being relatively new to civil engineering applications are not yet included in many construction codes. Thus, a municipal decision-maker who takes a strict interpretation of the code may simply reject the use of composite materials in the application. A less strict decision-maker may allow an exception for using composite as long as equivalent performance can be demonstrated. Either way, no follow on activity occurs in the composites value chain if the municipality decides that composites will not, or cannot, be used in the application. The end-user is key to the entire business development process and must be convinced that using composite materials is the best way to meet their needs. Other value chain players will follow once the end-user decides to use composite for their application.

Purchasing and Engineering Organizations Favor Incumbent Materials

Large organizations are driven to optimize their current products and incumbent material solutions. Industries that use large amounts of a given commodity tend to specialize in the purchasing of that commodity or product category in order to squeeze every penny of value to the company's bottom line. For example, steel components account for about 95% of the material cost of a typical washing machine [3]. Plastics account for the remaining 5% of the unit's material cost. A 10% reduction in the purchase price of sheet steel would have 19 times the impact on the final material cost of the washing machine versus a 10% reduction in plastic pricing. With this relative importance, management focuses its resources accordingly on finding ways to reduce the cost of sheet steel rather than plastic in its final product (see Figure 10.4). The washing machine company with its limited resources may even divert purchasing managers away from buying plastics toward finding increasingly cheaper sources of steel. This is rational. However when a new opportunity concerning composite materials arises (a form of plastics in the eyes of many customers), management may be unconvinced that a cost reduction, or any activity, concerning plastics can be as impactful as attention paid to its steel category. Extraordinary effort is required on the part of the business developer to show how the use of composites can have a meaningful impact on the overall cost of the washing machine.

Similarly, engineers are directed to devote their efforts toward optimizing the performance of their incumbent steel designs before considering a fundamental change from metal to composite. Dedication to continuous improvement becomes an impediment to introducing material alternatives. Successful improvements to the incumbent material raise the performance levels that the composite material must exceed in order to be worth the change. Consider the example of Japanese auto manufacturers. Simply put, they use fewer composite materials in their vehicles than

Typical Washing Machine Material Cost Breakdown

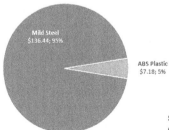

Material	Base Cost	Cost with 10% Reduction in Steel	Cost with 10% Reduction in Plastic
Mild Steel	$136.44	$122.80	$136.44
ABS Plastic	$7.18	$7.18	$6.46
Total	$143.62	$129.98	$142.90

Steel commodity receives more attention from Purchasing Department due to larger potential cost savings impact.

FIGURE 10.4 Washing machine cost breakdown according to material type. Typical focus of purchasing/engineering departments directed to larger, incumbent material categories.

American OEMs because they have done a better job optimizing their steel designs for lightweighting [4].

For the business developer to overcome this challenge, she must assemble an extremely convincing business case to garner interest from the customer. The business developer must appreciate that a customer's resistance to change often goes beyond competition against the incumbent material on a specific application. There is competition against the incumbent material as a complete category within the customer's organization. The savings on a given application must be large enough, the weight savings great enough, and the performance benefit important enough to capture the attention of the customer – broadly speaking. It helps if the customer has a high-level initiative such as an overall weight savings or sustainability target. A bio-based composite, for example, could have special appeal and receive strategic support for its use. Life cycle costing (LCC) and life cycle analyses (LCA), both favorable to composites, are growing in their acceptance by customers as factors to be included in the business case.

DIVERGENT METRICS, COMPLEX DECISION-MAKING

Another challenge occurs when decision-makers within organizations have disparate and often conflicting interests based upon different performance metrics and accounting practices. Engineers are rewarded when their products work according to customer expectations both initially and over time. New products are expected to meet critical engineering specifications, be delivered on time, and not exceeded accounting budgets associated with product development. Purchasing managers are assigned to source the lowest cost suppliers who are able to meet the engineer's requirements at a level that is sufficient – and no more.

While the engineer's metrics may extend for years after start of production, purchasing managers usually involve themselves only in the final stages of development. Purchasing managers are measured on an annual basis often without regard for the product life cycle. Management places greater value on short-term savings rather than savings over the long run. The longer the product life cycle, the greater the separation between purchasing's and engineering's metrics. The purchasing managers see the financial results of their purchase decisions almost right away. However, the engineering consequences of purchasing poor quality parts may not reveal themselves until years later.

How does this impact composites business development? Savings for composites are usually realized in the "use phase." However, they often come with a higher upfront cost. What is a lower total cost solution for an OEM end-user can appear more expensive in the beginning, but saves cost in the end. Composites, for example, do not corrode. Therefore, they require less maintenance or secondary treatments. Composites may have higher piece costs but are less expensive from a tooling perspective. Composite tooling can be more expensive than die cast tooling, but die cast tooling must be replaced several times over the production life of the product [5]. Frequently, only the initial tooling cost is factored into the OEM's business case. Replacement tool cost is assigned to a "maintenance" budget. Decisions can be unexpected or appear irrational if the business developer does not appreciate

I.   Proposal Description

II.   Marketing Rationale (customer impact)

III.   Volume Impact / Forecast

IV.   Engineering Impact (design, TRL*, weight, etc.)

V.   Manufacturing Impact (if "make")

VI.   Purchasing Impact (if "buy")

VII.   Assembly Impact (changes, line speed, complexity, balance, etc.)

VIII.   Service Impact

IX.   Warranty/Quality Impact (ties to TRL)

X.   Financial Roll-Up (piece cost, tooling, investment)

FIGURE 10.5   OEM business case outline.

the customer's method of accounting. Engineers and purchasing managers are very aware of their competing objectives. However, the problem persists because neither owns the issue from an organizational perspective. Composites business developers must carefully navigate the complex and sometimes counterintuitive accounting systems of large organizations. She must find opportunities that can satisfy each stakeholders' interests and may need to develop separate business cases that frame the benefits of the proposal from several viewpoints.

In addition, the highest person on the organization chart does not always make the decisions. Rather, it is a combination of persons within the organization. Miller Heiman's Strategic Selling [6] and John Asher's, Close Deals Faster [7] provide excellent guidance concerning strategic account management and long-term business development. In combination with how an OEM makes a decision, the criteria for making such a decision can be put into common categories. The below list is an example used by one OEM to consider its new product ideas. Answers to the questions constituted its standard business case (Figure 10.5).

CHALLENGES AT THE MOLDER

End-users do not do their own composites part molding unless vertically integrated. Usually, parts are purchased from a composites molder leaving the end-user (OEM) to focus on assembly. The composites molder must be convinced by the business developer that the composite material is in their best interest along with that of the end-user/OEM. If the end-user is already sold on using composite, then this lessens the work for the molder. However, the question remains whether there is a value proposition for the molder itself. The molder has fewer motivations than the end-user. Some of the molder's considerations include:

- Will the new product improve the business of their customer?
- Will the new product fill a void in the general market or simply cannibalize an existing product?
- Will the company itself increase its revenues, profits?
- Is the molder an incumbent supplier to the OEM with conflicting investments/interests in technology, i.e., will the new product obsolete current

Composites New Business Development Toolkit

☐ **Company and New Product Introductory Presentations**
  * Company overview
  * Existing engagements with customer
  * Overall composite product offerings relevant to customer
  * Application examples including business case
  * Capabilities and collaboration examples

☐ **Physical Composite Part Examples**

☐ **Levels I – III Material Property Data and Design Guide**
  * (I)   Basic Sell Sheets, Technical Data Sheets, MSDS
  * (II)   FEA Material Card
  * (III)   Specific design tips and recommendations

☐ **Customer-specific Development Proposal and Resources**

FIGURE 10.6   Elements of a composites new business development "Toolkit."

investments or will the composite material work in the molder's existing equipment?
* Will capital investment be required? If so, does the investment align with the high level strategy of the molding company?
* Will the molder be competing against an incumbent supplier using traditional materials? What happens if the incumbent supplier simply lowers their price?
* Does the new product improve operating efficiencies, address a new regulation, or alleviate an environmental concern?

The molder itself must understand the value proposition vis-à-vis the OEM: what is driving the OEM to use composite in the particular application? What is the incumbent material or next best alternative? Does the molder have a sales package for continuing to sell the OEM on the new technology for other applications? At a minimum, answers to these questions should be included in a new business developer's composite "toolkit." The toolkit should include sell sheets, technical data sheets on the new composite material, design and process guidelines, material safety data sheets, molded product samples, and more (see Figure 10.6). The requirement for more detailed information corresponds with advancement through the customer's Stage-Gate process.

CHALLENGES AT THE RAW MATERIAL SUPPLIER

The raw material supplier sells her product to the composites molder in most cases. The raw material supplier has the most difficult job to align its product with the needs of the end-user and all points in between. The business developer's role is to

I.   Opportunity Description and Program Goal
II.   Market Environment including Value Chain
III.   Strategy and Program Tactics
IV.   Product Description
V.   Product Features and Benefits
VI.   Customer Value Propositions
VII.   Business Case and Customer Cost Model including LCC/LCA
VIII. Product Pricing and Positioning
IX.   Comparison to Competition
X.   Environmental, Health and Safety Topics
XI.   Available Sales Tools and Supports, How to Order
XII.   Frequently Asked Questions

FIGURE 10.7   The composites Product Dossier™ table of contents.

furnish her sales team with the tools necessary to sell the value of their resin, reinforcement, or other product from which to mold the part. Business developers can provide its sales teams with what is called a Product Dossier™ whenever the company introduces a new product to the market. The Product Dossier is very effective in providing a single document that explains the rationale for the product, the product's value proposition along the supply chain, marketing collateral available to the salesperson, and more. Figure 10.7 provides an outline of the Product Dossier table of contents. Note while the new business development toolkit is external information for the customer, the Product Dossier is an internal document for salespersons at the raw material supplier.

NEW BUSINESS DEVELOPMENT EXAMPLES

The following examples illustrate some of the important considerations for successful composite business development. Each presents a context with different challenges.

AN EXISTING PRODUCT INTO A NEW REGION

A global manufacturer of building materials had success in offering one of its accessory building products in Canada. The product was a specially engineered polyethylene housewrap intended to replace organic felt in the home building process. Traditional paper felts had several disadvantages. They tore easily during installation, especially when wet. They had a tendency to rot and become less effective over time. Treatments used to inhibit bulk water intrusion also tended to trap moisture and promote the growth of mold. A new polymer composite housewrap eliminated these problems in addition to being lightweight. The company sought to expand the success of its new composite housewrap with homebuilders in the United States.

The business development challenge involved preparing the company's sales force for introducing the product to the US market with appropriate pricing and against local competition. The company's business case considered whether to import the product from Canada or manufacture it in the United States. This comparison had to consider foreign exchange rates, import duties and taxes, safety declarations, and cross-border logistics. Customer needs were verified as similar to, but not the same as, those in Canada since there are significant differences in home building codes across the United States. Pricing was analyzed and set according to the local market conditions. The company considered all of these aspects in its rollout strategy. From there, it created a Product Dossier for the company sales team to educate itself and to support the US product launch.

The company included all the elements in the Product Dossier outlined in Figure 10.7. In this case, the manufacturer decided on a strategy that promoted the convenience of offering the new composite housewrap product as a "top load" or add-on sale to its existing customer orders (see Figures 10.8 and 10.9). The manufacturer succeeded with the new product introduction and grew its category sales and profits by more than 10% in its first year.

A New Technology into an Existing Market

A manufacturer of a popular sports car expressed to its tier one supplier a desire to convert the vehicle's cast magnesium roof frame to an even lighter weight epoxy carbon fiber composite design using advances in the resin transfer molding process. Motivations for converting the magnesium roof frame to a carbon fiber composites design included those shown in Table 10.2.

An agreement to undertake the project was established among the OEM, tier one molder, and thermoset resin system supplier. The OEM provided cost and performance targets. The tier one and resin supplier formed an agreement to comanage the project with each sharing costs, engineering knowledge, and technical expertise. Fundamental to the success of the project was use of a newly developed family of fast

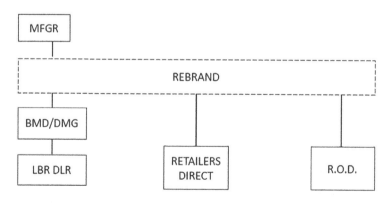

FIGURE 10.8 Value chain from Product Dossier – building material case example.

DISTRIBUTOR VALUE PROPOSITION

- TOP LOADING ROOFING, INSULATION AND FOAM
  - ✓ Handle only paper
  - ✓ One phone call
  - ✓ Improved turns
  - ✓ Easy brokerage profits
- USER FRIENDLY PACKAGE
- BUYING EFFIECIENCY – ONE VENDOR
  FOR FOUR PRODUCTS
- PULL THROUGH PROGRAMS
- COMPETITIVE PRICE
- EASE OF AVAILABILTY

RETAILER VALUE PROPOSITION

- COMPETITIVE PRICE
- TOP LOADING ROOFING, INSULATION AND FOAM
  - ✓ Efficiencies – delivering, unloading
  - ✓ One phone call/sales call
  - ✓ Turn improvement
- PACKAGING
  - ✓ Right size
  - ✓ Protection all the way to the job
- COLOR TIE-IN
- ENERGY EFFICIENCY BRAND NAME
- BUYING EFFICIENCIES – ONE VENDOR
- FOR FOUR PRODUCTS
- FULL PULL THROUGH PROGRAMS
- MERCHANDISING SUPPORT

FIGURE 10.9 Value chain and descriptive value proposition from Product Dossier building material case example.

TABLE 10.2

Features and Benefits of Converting a Magnesium Roof Frame to Carbon Fiber Reinforced (CFRP) Composite

Feature/Attribute	Benefit
Reduced part weight	Easier for on/off customer handling
Improved part stiffness and dimensional stability	Eliminates "hand finessing" production line issues frequently encountered with cast magnesium part
Reduced weight from a high location on the car	Lower vehicle center of gravity for improved speed, performance, and fuel economy
Roof frame design that reduces an excessive bond gap on the CFRP roof option	Improved performance and reduced cost due to less use of adhesive
Roof frame geometry is ideal candidate for the high-pressure resin transfer molding (HP-RTM) process	HP-RTM/LCM technology enables high volume production (e.g., >50,000 p.a.) of the roof frame as well as other composite part designs
HP-RTM is rapid and proven technology for producing epoxy CFRP components	Eliminates safety and supply chain concerns related to die cast magnesium parts

curing resins capable of supporting high-volume production programs (e.g., >50,000 parts per year). The tier one and resin company recruited additional partners to the project team based on each company's real-world experience with the new process. The supplier companies provided expertise in project management, resin and binder technology, reinforcements, part design and analysis, preforming, tool design, process simulation, production, and assembly system designs. In all, ten supplier companies plus the OEM participated to the project.

Experience dictated that it was important to establish how the part would be constructed from the preforming perspective. This would be the major cost driver for

TABLE 10.3

Features and Benefits of Converting Magnesium Roof Frame to CFRP Composite

Preforming Approach	Advantages	Disadvantages
"Donut Hole" (preform cut from single plies)	• No overlapping joints • Less likelihood of "misaligned" preform • Faster preforming	• Large (costly) area of material cut from center of frame • Potentially difficult to form fabric to contours
"Bacon Strip" (sides assembled to make preform)	• More efficient use of material; material used to construct sides of frame only	• Overlapping joints must be created to form frame • More assembly steps • Greater likelihood of "misaligned" preforms

the project. The geometry of the roof frame called for one of two general approaches for assembling the preform. The first approach was to cut a "hole" from a single fabric stack, leaving the sides and then molding the frame. The other approach was to assemble the individual "sides" of the frame and then mold the frame. The team referred to these approaches as the "donut hole" versus "bacon strip" designs, respectively. Each approach had its advantages and disadvantages (see Table 10.3).

The team's comparison concluded that the "bacon strip" preforming approach was more economical than the "donut hole" approach despite the higher likelihood of misalignments or scrap when creating the frame joints. Scrap material cut from the center of the donut hole approach also confirmed to be too great to offset from a cost perspective. Figure 10.10 shows the comparison of the two preforming approaches and thus the design direction for the project. The team determined that it could save an estimated 26% on materials and processing by using the "bacon strip" approach. As part of an overall business case, this type of analysis is important for determining feasibility and the most cost-effective engineering approach for the project.

The cross-functional team met weekly for nearly 2 years solving various design problems, performing process simulations, and conducting physical tests. The team commissioned a tool and successfully produced a series of prototype parts. The key to the development was having representatives from each link in the supply chain, each contributing expertise, and an upfront agreement among the key players involved. The final version of the carbon fiber roof frame was successful in meeting the performance criteria of the incumbent magnesium roof frame design while reducing weight:

- Redesigned lightweight carbon fiber roof frame vs. original die cast magnesium
  - CFRP = 1.8 kg vs. magnesium = 3.2 kg including hardware
  - A 44% weight savings on the roof frame itself; 32% weight savings on the overall assembly

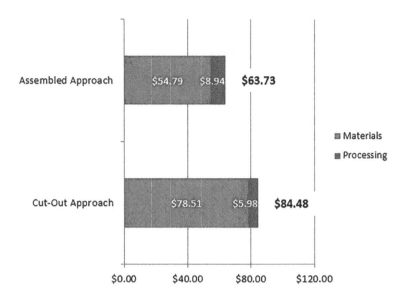

FIGURE 10.10   Cost comparison of different CFRP preforming approaches.

- Prototypes successfully produced on production scale high-pressure resin transfer molding (HP-RTM) equipment with a demonstrated cycle time of 3 minutes per part – unoptimized
- Met OEM roof frame static stiffness specifications
- Passed OEM dynamic side pole and frontal impact tests
- Completed HP-RTM material qualification plan transferable to similar parts and applications

In the end, the project showed the successful collaboration by several expert supply chain partners to develop a lightweight composite part made using the high-pressure resin transfer molding process. The experience and lessons learned were applied to similar projects [8].

An Existing Product into a New International Market

In the early 2000s, a composite resin manufacturer anticipated the growth of the China market for its resin products and started construction of a new reactor. Beyond a few anchor customers, the company began business development efforts to fill its plant capacity. The wind and automotive segments performed well for the company in its other regions. The company leveraged its qualifications with those customers who also planned to establish production in China.

The company's goal was to duplicate its resin products made in Europe and North America. While meeting such a goal is easier than designing resins from scratch, transferring an existing resin formula to a new region for production is very difficult too. Challenges include the fact that equipment differs, local raw materials are different, workers are not the same, and sometimes even process steps must be different. Nonetheless, the final product must be identical in performance.

Local market dynamics contrasted greatly between China and the West. Customer needs were similar but differed with respect to the value placed upon technical service and marketing supports. Customers were less mature in their understanding of composite technology. Therefore, Chinese customers tended to value brand affiliations and low pricing even if the true cost of an inferior product produced higher scrap in their plants. The company was used to selling differentiated, high value resin systems. It had to learn quickly that it could only capture the value of basic products with its Chinese customers. The company concluded that its premium products would not sell until its basic products were established in the local market.

The Chinese market for composites technology has matured since the 2000s. Local players advanced quickly. Producers are more sophisticated and business is fast paced. Western company business developers need to adjust their efforts according to the local market in order to be successful.

SUMMARY

The business development challenges for composite materials are many. Since each circumstance is different, solutions vary, but there are common themes. Developers should be prepared for the obvious and the not so obvious. They must be ready to educate their customers, consider the length and complexity of the composites supply chain, know that incumbent materials have structural advantages, and recognize that metrics differ within large organizations. These are thought starters for avoiding important pitfalls. If these obstacles can be overcome, the future for composites growth is bright.

REFERENCES

1. *Product Leadership: Creating and Launching Superior New Products, Product Leadership*, R.G. Cooper, Basic Books, Oakville, Ontario, August 18, 1999.
2. *Survey: Barriers to Implementing Lightweight Technologies*, Center for Automotive Research, https://www.cargroup.org, Ann Arbor, MI, 2018.
3. American Iron and Steel Institute, https://www.steel.org/steel-markets/appliances/, Washington, DC, 2020.
4. Automotive Composites Alliance (ACA)/American Composites Manufacturers Association (ACMA)Parts List, Detroit, MI, 2007.
5. Recent case studies of engineering thermosets for under-the-hood applications, C.A. Ball et al., *Society of Automotive Engineers Technical Paper*, Novi, MI, September 2013.
6. *Strategic Selling: The Unique Sales System Proven Successful by America's Best Companies*, R.B. Miller, S.E. Heiman, et al. Grand Central Publishing, New York, April 20, 2005.
7. *Close Deals Faster*, J. Asher, Ideapress Publishing, https://www.ideapresspublishing. com/, January 10, 2018.
8. Development of an Epoxy carbon fiber reinforced roof frame using the high pressure resin transfer molding (HP-RTM) process, C.A. Ball et al., *Society of Automotive Engineers Technical Paper*, 2020-01-0773, Novi, MI, April 2020.

11 Composites in Automotive through the Years

Dave Reed

CONTENTS

Pictures obtained from https://www.moldedfiberglass.com/about-us/history/corvette-story/

The developments of modern composites materials and processes have been closely linked to the development of modern cars and trucks. Since the 1950s, automotive engineers have demanded lower mass and lower cost alternatives to metal components often integrating multiple features for added value. Time and again plastic composites and processes have been developed to provide the solutions for these challenging applications. Shown below is a list of some of the production vehicles that have used composites for most of the body and their total production volumes.

1953–1954 Kaiser Darrin 425 vehicles [1]
1953–2021 Chevrolet Corvette, Cadillac XLR over 1,750,000 vehicles [2]
1957–2021 Lotus Elite1 Elan, Europa, Elan2, Elite75, Esprit, Elise, Excel, Elise, Evora, Exige, Opel Speedster total est. over 112,000 vehicles [3]
1973–1983 Matra Bagherra, Murena over 58,700 vehicles [4,5]
1984–2002 Renault Matra Espace Avantime mini-van over 367,000 vehicles [6]

DOI: 10.1201/9781003161738-11

1983–1988 Pontiac Fiero, SE, GT 370,168 vehicles [7]
1990–2007 Saturn (notable, but most panels are thermoplastic not composites) [8]
1990–1996 Chevrolet Lumina APV, Pontiac Transport, Olds Silhouette [9]
1992–2016 Dodge Viper GTS, RT10, ACR 31,850 vehicles [10]
1997–1999 General Motors EV1 Electric Vehicle 1117 vehicles [11]
1997–2021 McLaren F1, MP412C, 600LT, P1, 650, 720S over 25,000 vehicles [12]
1998–2002 Camaro-Z28 and SS, Firebird-Trans Am, WS-6 (steel rear fenders) [13]
2001–2021 Aston Martin Vanquish1&2 DBS1&2 (carbon) over 15,000 vehicles [14]
2002–2005 Ford Thunderbird (steel rear fenders & doors) over 68,000 vehicles [15]
2005–2007 Mercedes Benz – McLaren SLR (carbon fiber epoxy) 1151 vehicles [16]

*1988 Fiero GT

*2002 Camaro SS

*2006 Corvette

*2021 Corvette

Composites in heavy-duty truck hoods, cabs, and sleepers are among the most massive parts as well.
*Courtesy Dave Reed, Author & Owner – Photos of Composite Cars

This list of composite automotive bodies brings up a question, why some cars choose composite bodies and not others? Several factors must be considered to choose composites over steel or aluminum – mass, production volume, material cost, processing cost/cycle time, and tooling cost/lead time.

High-performance cars place greater value on saving mass, so composites or aluminum are preferred. High-performance cars are usually lower production volume, so the lower cost of composites tooling becomes more important, and the longer cycle times of composites molding is less of an issue. High-performance cars often choose composites to allow more complex shapes and to mold in aero shapes and air scoops without add-on parts. Customers of high-performance cars also demand more up-to-date styling and more frequent model changes which are easier with lower cost composites tooling. Another important factor in choosing composites is the Make-or-Buy decision. Many automotive companies have large investments in expensive stamping presses, welding equipment, paint systems, and plants. This "sunk capital" and stamping, welding, and painting experience often pushes the decision to steel or aluminum instead of composites and avoids out-sourcing the body parts to a composites molder.

A brief history of the Corvette body composites represents much of the continuing advances in composite body materials and processes throughout the automotive industry. Composites were chosen for the first-generation Corvette body for several reasons still important today – design flexibility, customer excitement, low mass, damage resistance, corrosion resistance, fast tooling time, and low tooling cost.

The first-generation C1 Corvette 1953–1962 presented a major challenge and opportunity for automotive composites. The decision to start production in the summer of 1953 was not made until the Corvette show cars drew enthusiastic crowds and media attention at the 1953 winter car shows in New York City. With only a few months to start of production set for June 1953, the short tooling time and low tooling cost of composites along with lighter weight were major advantages over the familiar steel stamping and welding processes. The 1953 production model run was only 300 cars all molded by the Molded Fiber Glass Company® in Ashtabula Ohio. The MFG Company®, started by Robert Morrison, pioneered the use of composites in automotive. Mr. Morrison chose to mold the Corvette body panels with the revolutionary matched metal die process instead of the conventional hand layup process. The completed composite bodies were then assembled with the chassis, powertrain, and body interior on a Chevrolet preproduction assembly line in Flint Michigan. To expedite production start-up, all 1953 Corvettes were painted white with red interiors. In 1954, production was increased tenfold and moved to the St Louis Chevrolet Assembly plant. However, many challenges of fit and finish of the FRP needed to be addressed for automotive quality and production rates. All the part edges had to be hand trimmed and each appearance surface had to be sanded and repainted to reduce surface imperfections. The MFG Companies® has continued to innovate and improve on many of these challenges under the leadership of Richard Morrison.

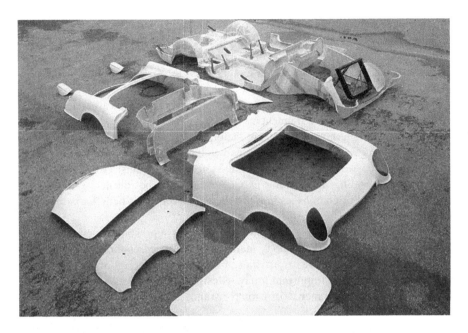

Pictures Obtained from – https://www.moldedfiberglass.com/about-us/history/corvette-story/.

Production rates jumped to 3,467 vehicles by 1956 with new styling and popular options like the new V8 engine and 4 speed manual transmission. The increasing production rates meant the hand finishing became an increasing challenge for the C1 and C2 Corvettes.

The C2 Corvette from 1963 to 1967 was made by the same matched metal die molding process, and the production climbed to 21,513, but the improving finish made other surface defects more apparent. The fiberglass was not evenly distributed and would cause waviness in the surface. Also, the polyester adhesive used to bond the inner panels to the backside of the outer panels would cause a bond-line read-through condition from the polyester adhesive exotherm and localized thermal expansion. This bond-line appeared as a continuous wave in the finish painted side of the body panels and presented a new challenge. St. Louis also developed a technique of reducing paint defects by reflowing the fresh acrylic lacquer paint with a heat gun.

The C3 Corvette 1968–1982 introduced a new process called SMC molding. This process has become an industry standard because of its improved fit and finish cycle time and process consistency over earlier processes – important as production had climbed to a high of 53,807 in 1979. The new SMC composite also included an acrylic thermoplastic ingredient to reduce the surface waviness. The acrylic modifier, called Low-Profile, was intended to migrate to the surface of the SMC during molding to reduce the short-term waviness caused by the uneven distribution of the chopped fiberglass strands. Also, a new two-part urethane adhesive from The Goodyear Tire and Rubber Company®, then PPG Industries, Inc.®, and later by Ashland Global Holdings, Inc.® was introduced with the C3 to reduce the exotherm and bond-line read-through. The urethane adhesive did not create as much heat when it cured as the old polyester adhesive and it also provided a stronger tougher bond. However, early laboratory tests appeared to show the new urethane adhesive had wider variation in adhesion to the new SMC formulation. But closer examination revealed that the acrylic thermoplastic surface layer was being pulled off the new SMC composite by the stronger urethane adhesive. The short-term fix was to locally sand the bond area to remove the thin acrylic-rich layer before applying the urethane adhesive. The longer-term fix was to use an adhesive primer and modify the acrylic thermoplastic ingredient to increase the strength of the acrylic layer. Another improvement to the surface finish was to change the paint primer to a new urethane primer called "Polane"® from Sherwin Williams Company®. The new urethane paint primer had a slower cure rate that allowed much of the outgassing from surface porosity in the SMC to bubble through before the primer skinned over. This allowed the primer to flow back into the surface pores and seal them as it cured in the paint oven. An additional change at considerable expense was made for the 1978 model to have the SMC part molders apply the new urethane primer instead of priming at the St. Louis assembly plant. Having the molders apply the primer also gave them some quick feedback to tell them when their SMC processing or molding conditions were causing more surface porosity in the molded parts. During the 1982 model, Corvette production was moved from the St. Louis Assembly Plant in Tennessee to the new Bowling Green Assembly Plant in Kentucky dedicated just for Corvettes. The Bowling Green paint system was designed for the new two-part acrylic urethane base/clear enamel paint which further reduced paint defects and reduced paint solvent emissions for the plant.

The C4 Corvettes 1984–1996 benefited from a new In-Mold-Coating polymer layer on the appearance surface that was injected directly into the mold after the part was molded and before the part was removed from the mold. The In-Mold-Coating process was developed by the GM® Manufacturing Development Composites Department and quickly adopted by the Corvette body molders. This added layer sealed over any porosity on the appearance surface, but increased mold cycle time and piece costs. The C4 also benefited from a total redesign that integrated the hood with the top of the front fenders and created wider angles between the horizontal body surfaces and the vertical surfaces. This allowed more direct pressure during molding to reduce porosity on the molded panels as they curved from the horizontal toward the vertical. Styling integrated these modifications into a more aerodynamic body for lower wind resistance and a more modern shape. Customers liked the more rounded aerodynamic body profile and could now open their larger hood for full engine access and front suspension visibility. This new aerodynamic body shape also eliminated the C3 troublesome bonded joint between the horizontal panels and the vertical panels on the fenders. Those long horizontal joints on the C3 had required careful sanding and pit filling at the assembly plant before going to the paint line.

The C5 Corvette for 1997–2004 introduced a new SMC with a toughened modified polyester developed by molder ThyssenKrupp/Budd Division® that reduced the old porosity and surface cracking and short-term waviness so much that the In-Mold-Coating step was no longer required. The new SMC formulation was called TCA® for Tough Class A® SMC. This also allowed the panels to be made thinner. A new epoxy two-part adhesive from Lord Corporation® was also introduced to replace the urethane adhesive and further reduce the exotherm that caused bond-line read-through where the inner stiffening panels were bonded to the outer appearance panels. Also, MFG Companies® developed and introduced a carbon fiber epoxy laminate with balsa core for the floor pans for significant mass savings.

The C6 Corvette from 2005 to 2013 benefited from all of the previous composites improvements. Also, carbon fiber epoxy composites were introduced on the ZO6 and ZR1 models front fenders, hood, and roof to save more mass. These parts were molded by Plasan Carbon Composites Inc.® in a proprietary process that avoided laborious vacuum bag molding and reduced the molding time from hours to minutes.

The C7 Corvette from 2014 to 2019 used a faster cure carbon fiber epoxy composite in an SMC/prepreg process for the hood, roof, and front fenders. A new toughened polyester resin SMC replacing the calcium carbonate reinforcement with much lighter hollow glass bubbles was used for the rest of the body panels. This breakthrough called TCA Ultralite® SMC was developed by Continental Structural Plastics®, CSP®, and now Teijin CSP®.

The C8 2020 Corvette expanded the TCA® Ultralite SMC® to most of the body panels to further reduce mass. These continuous improvements over the decades eliminated the porosity and reduced scrap and rework, reduced the density of SMC from 1.9 to 1.2 spgr, reduced typical panel thickness from 4 mm to less than 3 mm, and reduced the cycle time from 3 to 4 minutes to about 2 minutes depending on part size and complexity. MFG® also introduced an even lower density SMC for nonappearance panels called Float® with a density less than 1.

Discussion of the Corvette composites would be incomplete without mentioning the innovative filament wound fiberglass epoxy transverse leaf suspension springs. The C2 had introduced an independent rear suspension with a steel transverse multi-leaf spring that significantly improved handling and provided more interior trunk room. This carried over into the C3 Corvette at 22 kg but was soon replaced with a composite mono-leaf transverse spring at just 3 kg which also eliminated corrosion and increased the durability [17]. This innovation was so successful it was used for both the front and rear transverse leaf springs on the C4, C5, C6, and C7 Corvettes. The C8 switched to coil springs front and rear due to mid-engine packaging needs.

The innovative mid-engine Pontiac Fiero sports car 1983–1988 also used composites to great advantage in the body. The horizontal panels – hood, roof, rear deck, and engine cover – were all SMC similar to the C4 Corvette composites at the time. And, the fenders and door outer panels were molded with a new composite of milled glass fibers and glass flakes in thermoset urethane by a process called reinforced reaction injection molding (RRIM). This innovative material and process were much lighter and more damage resistant than SMC and improved the surface finish too. The RRIM process also had a faster cycle time better suited to the Fiero higher production rates. Another successful composite innovation on the Fiero was the glass mat thermoplastic (GMT) polypropylene front trunk. This composite allowed the deep front trunk to be molded in black color in one piece rather than the multiple pieces that would have been required for SMC or steel stamping. This avoided welding and sealing the steel stampings together or bonding the SMC pieces which saved money and mass. This also avoided the need for corrosion protection or paint. This same process has also been used to form some automotive bumper beams, seat backs, load floors, the GM EV1 battery tray, and other large composite parts requiring exceptional toughness and strength.

There have been many decades of progress and many successes in automotive composites, but who was first to design and build a composite car and open the eyes of the automotive industry to the possibilities of composites? Henry Ford, the automotive pioneer himself, may have been the first to design and build a car with a composite body in 1941. Unfortunately, his vision was interrupted by the Second World War and the urgent need to get back into production at full volume with proven steel stampings and equipment as soon as the war ended. The following is a summary of the Ford Motor Company® composite car and photos from the henryford.com website.

The frame was made from tubular steel and had 14 composite panels attached to it. The car weighed 2,000 pounds, 1,000 pounds lighter than a steel car. One article claims the composite included soybeans, wheat, hemp, and ramie, while the man instrumental in creating the car, Lowell E. Overly, said it was "soybean fiber with formaldehyde used in the impregnation." Henry Ford wanted to combine the fruits of industry and agriculture and show that plastic panels made the car safer than steel and could even roll over without being crushed. Another reason was due to a shortage of metal at the time, he hoped his new composite might replace steel.

Pictures and history courtesy of The Henry Ford from henryford.org/collections-and-research.com website

REFERENCES

1. en.m.wikipedia.org/wiki/Kaiser_Darrin.
2. https://en.wikipedia.org/wiki/Chevrolet_Corvette#Production.
3. https://lotusltd.com/resources/lotus-cars-number/.
4. en.m.wikipedia.org/wiki/Matra_Bagheera.
5. en.m.wikipedia.org/wiki/Talbot_Matra_Murena.
6. en.m.wikipedia.org/wiki/Matra_Espace.
7. en.m.wikipedia.org/wiki/Pontiac_Fiero.

8. en.m.wikipedia.org/wiki/Saturn_Corporation.
9. en.m.wikipedia.org/wiki/Chevrolet_Lumina_APV.
10. en.m.wikipedia.org/wiki/Dodge_Viper.
11. en.m.wikipedia.org/wiki/General_Motors_EV1.
12. en.m.wikipedia.org/wiki/McLaren_Cars.
13. en.m.wikipedia.org/wiki/Chevrolet_Camaro_(fourth-generation).
14. en.m.wikipedia.org/wiki/Aston_Martin_Vanquish.
15. en.m.wikipedia.org/wiki/Ford_Thunderbird_(eleventh generation).
16. en.m.wikipedia.org/wiki/Mercedes_Benz_SLR_McLaren.
17. McLellan, D. (2002). *Corvette from the Inside.* Cambridge, MA: Bentley Publishers. pp 86–87. ISBN 0-8376-0859-7.

12 Recycling of Polymer Matrix Composite Materials

Brian Pillay, Kristin N. Hardin, and Haibin Ning

CONTENTS

INTRODUCTION

Composites are widely accepted as a material that achieves sustainability through lightweighting and hence lowers fuel consumption by increasing efficiency of structures and products. Their long life spans and durability give them an advantage on their material counterparts/rivals with regard to sustainability. However, composites are not as widely recycled as metals, glass, plastics, etc. The basic problem lies in the heterogeneity of composites where two or more different material types become difficult to separate at end-of-life (EOL) stages. Typically, composites are fiber-reinforced polymer matrix composites (PMCs) containing a resin and a fiber component. Each of these components has their challenges. Resins are broken up into thermoplastics and thermosets where thermoplastics are heat reversible and can be remolded. On the contrary, thermoset-based materials are not heat reversible, once polymerized, and, thus, are considered nonrecyclable. The majority of resin-filled materials in landfills are contributed to thermosets as they are considered nonrecyclable by the majority of the composites industry. Over time, as composites sit in the

DOI: 10.1201/9781003161738-12

landfill, they produce landfill gases (LFGs). LFGs contain harmful gases and volatile organic compounds that can adversely affect public health and the environment [1]. Diverting composites (specifically thermoset-based composites) from the landfill has the potential to decrease these negative impacts on public health and the environment. Diverting fiber-reinforced composites has the potential to decrease these negative impacts as well as cost for composite manufacturers. Fiber-reinforced polymer composites (FRPCs) are increasing their footprint in many applications and sectors due to their low weight, durability, and enhanced performance attributes. The high specific mechanical properties of glass and carbon FRPCs make them highly attractive and abundant in various markets. Carbon fiber-reinforced polymer (CFRP) composites are generally applicable to higher-end markets such as aerospace, whereas glass fiber-reinforced polymer (GFRP) composites make up lower-end or commodity markets. However, with new technology and improved manufacturing processes, both carbon fiber and glass fiber (GF) utilization has become more cost competitive and find uses in all market sectors. With their many strengths and broad versatility, FRPCs are inherently less recyclable as compared to metals. Due to the higher cost of carbon fiber, higher-end markets can afford to invest in high-temperature reclamation processes like pyrolysis and the eventual resizing of fibers. However, GF is cheap enough and so abundant that commodity markets do not feel this urgency. Therefore, little to no recycling is done with glass-filled composite scrap. GFRP composites constitute the largest class of composites with a worldwide market of roughly 4–5 million tons per year [2] and are dominated by the transportation, aerospace, and defense segments [3]. Recycling such a highly used, abundant, versatile, and advantageous material could open new markets for the composites industry. Although a general cost analysis may show that buying raw GF or neat GFRP is cheaper than reclaiming the glass or trying to find a recycling solution for the secondary material source as a whole, disposing of GFRP has negative economic and environmental impacts.

The United States is typically known as the "throw-away society" by most other countries and this culture has placed extreme pressure on our landfills and other method of disposing waste.

North America alone has tripled the amount of waste generated as seen from the decrease in the number of readily available landfills; there were approximately 18,500 active US landfills available in 1979, approximately 3,100 active landfills in 2010, and approximately 1,900 by 2013 [4,5]. This decrease in available space is exacerbated by the increase in population and thus the amount of trash produced. It can also be related to the increase in technology advancement in general with increasing applications for composite materials and composite waste stream material adding to this problem. Public health and sustainability are both at risk with the increased use and need for landfills as landfills take up valuable space and can emit landfill gases LFGs. LFGs are harmful gases containing volatile organic compounds (VOCs), hazardous air pollutants (HAPs), etc., that vent slowly into the atmosphere and leech into the groundwater [1]. Not only have LFGs been linked to the greenhouse effect, but research has also linked them to cancer and respiratory diseases [6]. Landfilling can also harm the bottom line of the composites industry. Landfilling costs money. Tipping fees are levied by landfill operators as a means to maintain the

TABLE 12.1
Summary of Landfill Tipping Fees in the United States from 2003 to 2018 [7,8]

	State	USD/ton
2018 Lowest	Mississippi	24.75
2018 Highest	Alaska	151.19
	US average 2008	44.09
	US average 2013	49.78
	US average 2018	55.11

space as well as offset the opening and eventually closing of it [7]. Tipping fees in the United States currently range from 24.75 to 151.19 USD/ton as shown in Table 12.1.

Pressure from consumers and advocacy groups alike is surmounting in the areas of waste management and environmental friendliness [9]. Recent years have seen an increase in overall consumer perception of sustainability with companies marketing their "green" products and promises to customers. Consumer preferences coupled with long life spans, advances in technology that increase available composite applications, continued decrease in available landfill space, growing environmental consciousness, and legislative mandates and policies are all drivers for composite recycling.

Sustainability and all it entails are becoming an important component to maintaining the current business of the composites industry as well as an integral part of future growth. The trick is figuring out where to start as composite uses are far-reaching; consumer waste, electronic waste, the construction industry, the marine industry, etc. On a grand scale, this is where composite manufacturers can play a big role. Instead of targeting a single industry like the sporting goods or E-waste, one should target each composite product at the beginning of its life at the manufacturing facility. Not to mention, composite manufacturing can lead to a number of sources of scrap that invariably end up in landfills, which is costly and damaging to the environment. There are six main sources of scrap produced at a composites production facility: raw material scrap, raw material trim-offs, part trim-offs, machining and finishing scrap, reject parts, and EOL components. In 2014, an industry survey was produced to gauge where the composites industry stood with respect to recycling. The survey was sent to a cross section of American Composites Manufacturers Association (ACMA) member companies through the Recycling Committee of the ACMA Green Composites Council. Most manufacturers noted that 5%–10% of their total production volume was landfilled with some citing as high as 30% [10]. Combining the former finding with Moffit's finding that four billion pounds of resin and composites are produced annually [11] and assuming a minimum manufacturing scrap rate of 5% that constitutes approximately 200 million pounds landfilled each year. This number is only projected to increase as the use of composites has continued to increase over the past two decades. Combining this 5%–10% scrap rate with Lucintel's projections, the composite industry lost as much as $10.53 billion

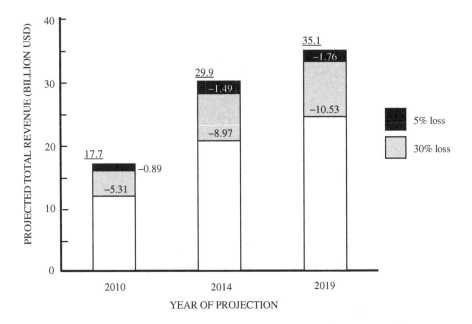

FIGURE 12.1 Total revenue for the composites industry per 2010, 2014, and 2019 as projected by the Global Composites Market [12] combined with ACMA survey results of scrap generation to show a possible projected loss in revenue [10].

in revenue in 2019, as shown in Figure 12.1 [7,12]. Sustainability in the composites industry must be addressed. There are currently some accepted methods for composite recycling, which are described in the following section. The survey also asked if participants knew their costs associated with landfilling, which over half did not. Those that did cite $0.05–$0.20 per pound [10]. The economic, let alone environmental, impact with regard to landfilling on the composites industry is grave.

CHALLENGES OF RECYCLING COMPOSITES

Technical and systemic challenges exist in composites recycling. Those challenges

TECHNICAL CHALLENGE

One of the most fundamental aspects of the successfully manufacturing composites is the ability to achieve good fiber–matrix bonding. This is applicable for all types of PMCs with any reinforcement, carbon, or GF. All fiber manufacturers commit significant financial resources to continually strive to improve the sizing chemistry for the different matrixes, both thermoset and thermoplastic, yielding higher mechanical properties. The sizing chemistries are closely guarded trade secrets by respective fiber manufacturers. Depending on the application sector a fiber manufacturer targets, the sizing chemistry may change even when the matrix may be the same. The sizing acts as a coupling agent between the fiber and matrix, and the ability to transfer load from the matrix to the fiber dictates the strength in the composite. There are thousands of sizing chemistry formulations applied to fibers in order to extract the highest properties for the specific applications.

Further to the fiber manufacturers, resin manufactures are constantly research formulations that could improve the bonding and wettability of fibers during the manufacturing process. The formulations are also modified to tailor viscosities, gel times, cure times, etc., to improve the manufacturability of the composite, depending on the processing methods. This is true for both thermoplastic and thermoset processes. The increase in the use of composites in the automotive sector has led to pressure to reduce cycle times for part production. This has been typically achieved by modifying manufacturing processes, which leads to modifying the chemistry of the matrix/resin.

The strong bond between the fiber and matrix makes it very difficult to recycle, as the resin and fiber cannot be easily separated in EOL components and production scrap. This is obviously a huge challenge as the competing need for strong bonds during use is paramount in composites which is one of the major hindrances to successfully recycling at the EOL.

Thermoplastic composite materials are said to be inherently recyclable as their chemistry allows them to be reheated to plasticity and remolded; thus, secondary thermoplastic composite scrap is easily reprocessed. If, however, the thermoplastic is reinforced with fibers or sheet materials, the process may require mechanical grinding or shredding to reduce the size. Traditional techniques to recycle thermoset composite materials also include size reduction techniques and thermal and chemical degradation. All of these techniques involve specific equipment that can drive up the cost for the recycling process. Mechanical reduction includes shredding, grinding, and/or granulating EOL composites for use as filler materials typically in lower property, i.e., down-cycled, applications than the original composite.

Chemical degradation works to remove the matrix by reverse polymerization with organic or inorganic solvents or even supercritical fluids. This method allows for a cleaner and less corrupted fiber but can generate potentially harmful waste chemicals [13]. In thermal recycling of thermoset composites, fiber reinforcements are separated from the polymer matrix by burning off the resin and recovering the fibers. This process can be useful especially when the recovered fibers are expensive like carbon. Thermal degradation can maximize the repurposing of the composite by breaking the waste into raw material and eventually energy [13–16]. The principle of this technique is to heat the composite to a high enough temperature to vaporize the resin leaving behind the reinforcements. Further the condensation of the volatiles creates fuel and can self-sustain the burning process.

Two techniques are used under thermal recycling. The first being a fluidized bed thermal process and the second is the pyrolysis process (used for carbon and GFs). Care should be taken as this method can damage the fibers due to shrinkage and produce thermal stresses. A fluidized bed thermal process heating at 450°C–650°C reduces tensile strength of GFs by 50% and 90%, respectively. For carbon at 550°C, Pickering observed 20% reduction in the tensile strength with stiffness intact [17]. Materials Innovation Technologies (MIT), Inc., a subsidiary of Carbon Conversions, Inc., has a reclamation and manufacturing unit that addresses composite sustainability by repurposing carbon fiber components in a means to close the loop on a product's life. MIT developed a process called 3-DEP where carbon fibers can be reclaimed from either manufacturing scrap or EOL parts to be utilized in the production of pre-forms and other high-end products. This process begins with pyrolysis and it is ideally suited for processing recycled or reclaimed fibers that are discontinuous. Once the resin has been burned off, the fibers are intimately mixed in the 3-DEP process. This process uses hydro-entanglement to develop a preform, which is then infused with resin using traditional processes like RTM. The reclamation of carbon fiber results in 96% less energy than the manufacturing of virgin carbon fiber [18].

SYSTEMIC CHALLENGES TO RECYCLING

The composites industry is very vast and diverse with respect to materials, market sectors, applications, materials used, etc. This poses a huge challenge in coordinating the activities of the various industry sectors. The industry does not provide pathways for recycling and individual companies that are too fragmented and do not have the scale of operations to include robust recycling programs. The obvious choice is to focus attention on minimizing scrap during manufacturing and ignore the problem at the EOL as the onus does not fall on the manufacturer to recycle or dispose of components. Minimizing scrap is always ideal; however, the nature of composites manufacturing always leads to some scrap generation and totally eliminating it is almost impossible.

The cost of landfilling, although steadily increasing in price and declining in availability, is still cheap in the United States compared to some of the other parts of the world like Europe and Japan. This does not provide enough of an incentive for companies to look for alternatives to recycling. Furthermore legislation in the United States is very fragmented with states deciding on mandates, or lack thereof, to incentivize companies to recycle rather than landfill. States on the West coast and Northeast lead the way with respect to environmental legislation compared to states mainly in the South, Southwest, and Midwest. The motivation for a consolidated recycling approach diminishes as the legalization is so vastly different, with respect to environmental and health and safety.

One company does not produce enough manufacturing waste to make recycling effective with respect to volume generation. The best scenario will be to reintroduce the material into the mainstream components which is very difficult to do and offers very limited possibilities. EOL components are also scattered and need to be collected and sorted. Depending on the manufacturer of the components, the materials used may be different with respect to formulation and additives. This further compounds the problem, as most time the manufacturer will not divulge the formulation of materials used, as its considered trade secret and they fear it will jeopardize their competitive advantage. Clearly there needs to be a coordinated effort to incentivize manufacturers to collaborate on a recycling path forward.

This will allow for more transparency in material formulation, potentially standardizing materials used in certain sectors with high volumes of EOL parts, like the automotive sector. This will allow for proper materials safety data sheets to be produced for the recycled materials. Proper safety protocols will be developed, so operators will be able to use the proper personal protective equipment when working with the materials. Standardization will also allow for recyclers to better monitor the quality of the recycled material produced, so users of the recycled material will be guaranteed that minimum standards are met and have confidence in using the materials. It will also alleviate the concern of ensuring that consistency of supply will be met by both manufacturing scrap and EOL components.

The plastics (although limited) or metals recycling industry can be used as a model to develop the infrastructure, collection agencies, transportation, storage, etc., to ensure that systems are in place to develop the supply chain needed for success. Potential end-users will have the confidence to build their business models and make investments accordingly. A large component of building the infrastructure includes education. Educating existing workers in the composites industry to treat all materials as a potential secondary application material that generates revenue rather than treating it as waste to be landfilled. The materials will need to be collected prior to being contaminated to get maximum value in the recycled stream.

CASE STUDIES OF COMPOSITES RECYCLING

There has been limited success achieved in recycling programs, a few are discussed below.

RECYCLING OF GF THERMOSET COMPOSITES VIA CEMENT KILNS

The Cement Kiln method of recycling combines mechanical reduction techniques and thermal degradation. Either from manufacturing waste or from EOL components, GF thermoset composite scrap is processed into a granulated form that is then used as either raw energy or a source of energy in cement manufacturing widely in Europe. Thermoset FRPC has a very high 10 calorific value of around 30,000 J/g [19]. This high-energy resource can be used as fuel. The problem is due to the high calorific value and the toxic fumes released that overload the incinerating systems, which increases costs. A solution to this problem can be burning them in cement kilns and replace part of clay and limestone with incombustible materials thus reducing the cost of the standard material. A 10% material replacement can be done without affecting the cement performance. At higher than 10% the boron oxide produced has a detrimental effect on cement performance [19]. The European Composites Industry Association (EuCIA) has reported that this route for recycling can reduce the carbon footprint of cement manufacturing by up to 16% [20].

AEROSPACE PREPREG SCRAP REUSE

Prepregs are very abundant in the aerospace industry. Researchers at the University of Southern California (USC) are taking advantage of their location to aerospace manufactures and have developed a prepreg scrap reuse program for nonprimary applications. Prepreg ply trim-off scrap is cut up into small "chips" and is used in bulk molding compounds (BMC) [21]. Studies provided by the team showed that these scrap prepreg-based parts can be competitive with other carbon fiber products due to the reduction in cost due to cuts in raw material costs as well as being competitive with traditional materials in low-cost applications where the use of carbon fiber was previously ignored due to higher raw material costs [21].

RECYCLING BY DESIGN

A more novel and life cycle assessment (LCA) type approach has been taken by Connora Technologies that uses chemical transformations of programmable cleavage groups in recyclable epoxy, modeled after thermoplastic epoxies [22]. While cross-links provide thermosets with mechanical advantages over their thermoplastic counterparts (i.e., polymers which are not cross-linked), they are also their Achilles heels that make them inherently nonrecyclable. Connora Technologies has been developing specially engineered amine hardeners, Recyclamines®, that work with any epoxy resin to create programmed thermosetting plastics. They found that for most applications, recyclable epoxy parallels the performance of conventional epoxy, yet it can be easily converted into a thermoplastic under a simple recycling process and thus surpass the popular qualities of conventional epoxies. As shown in Figure 12.2, the generic design of the recyclable epoxy hardener molecules is derived from a diamine structure, whereby the two amino end groups are tethered together by a central cleavable group [23]. As with conventional amine epoxy hardeners, the amino group reacts with epoxy resins. The generic tether is interchangeable and influences both processing properties (e.g., curing time and viscosity) of the precured system and mechanical properties (e.g., tensile strength, flexural strength, and glass transition temperature) of the final cured epoxy thermoset resin [22]. As the molecular structure of these special recyclable hardeners is akin to conventional amine hardeners, they also possess similar mechanical and processing properties. Further, the central programmed cleavage group is the key to the technology and is what makes the resultant thermoset "smart." Generic molecular design of recyclable epoxy hardeners used to create a programmed epoxy that can be converted into a well-defined thermoplastic [22].

The conditions required to induce cleavage of the crosslinks are temperatures of approximately 100°C and weak acid. Both the concentration and the pH of the acid required to induce the conversion into a thermoplastic are dictated by the chemical nature of the cleavage component. Polymers modified with acid-cleavable groups have been successfully implemented in both photoresist and drug delivery applications [24]. The common premise in these cases is that acid-induced cleavage will result in a solubility change in the parent material. The concept of recyclable epoxy technology is somewhat similar, albeit the desired change is to transform an otherwise intractable material into a tractable one that is of high inherent value. The cleavage groups in the programmed epoxy hardeners will exist at every cross-link point in the cured epoxy matrix. Immersion of the cured epoxy in a specific recycling bath with the previously mentioned parameters will prompt cleavage of the cross-links and conversion of the thermoset into its reclaimed thermoplastic epoxy counterpart [24]. This transformation occurs only upon immersion of the thermoset in a solution of sufficient temperature, pH, and concentration.

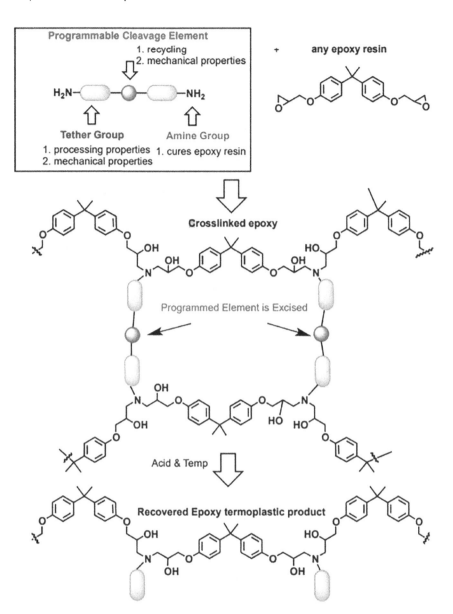

FIGURE 12.2 Generic molecular design of recyclable epoxy hardeners used to create a programmed epoxy that can be converted into a well-defined thermoplastic [22].

Generally, most epoxy composite applications operate in environments that are far from the parameters required to trigger conversion of the programmed epoxy. The thermoplastic offset is completely insoluble in water, but it will solubilize into the acidic recycling solution because of protonation of the polymer backbone. This makes it possible to recover any components of the epoxy composite such as fibers which may be physically removed from the solution. Subsequently, the dissolved epoxy is then recovered via an extra processing step. This novel technology allows composite manufacturers the fabrication of fully recyclable epoxy composites.

Compounding GF Epoxy/Vinyl Ester Composites with Thermoplastic Matrix

A study conducted at UAB showed that machining residue (MR) produced from pultruded machining GF epoxy composites could be compounded into thermoplastic resins like polypropylene (PP), nylon 6, etc., to improve the mechanical properties. The study included compounding cutoff from a GF vinyl ester composite product as well. The trim-off parts were reduced in size using a hammer mill to pulverize the composite into a fine powder. The GF epoxy composite MR was compounded with PP using a twin-screw extruder; the pellets were collected and passed through a single reciprocating screw plasticator; and a charge was collected, placed in a mold on a press, and compression molded into plates. The GF vinyl ester was compounded with nylon 6 and the same process was followed. In both cases, 30 wt% of the recycled material was blended into the neat polymer.

There was a 15% increase in tensile strength and 50% increase in flexural modulus of the PP recyclate compared to the neat PP. There was a 35% increase in tensile strength, 410% increase in elongation, and 13% decrease in tensile modulus for the nylon 6 recyclate compared to the neat nylon 6. It is clearly evident that there is significant potential to use the recycled thermoset materials as fillers in thermoplastic to improve mechanical properties.

Machining Residue of Nylon (PA) 66 GF Thermoplastic (LFT) Composite

The 750 clamps referred to in the case study in the chapter on Industry–University partnerships required post-process machining. Specifically four holes had to be machined on the flanges of the collar. As the collar was a significantly thick section, this very rapidly resulted in a large quantity of MR. This residue was collected and a study was performed on the ability to recycle PA66/GF composites. The material was collected in several large garbage bags directly off the machine, which included the cutting fluid used in the machining operation.

The collected MR was stored until the completion of the project (approximately 6 months) and the study was then conducted. This allowed for typical collection and timing at a composites manufacturing facility. The MR was then washed using two processes, acetone, soap, and water. The residue was analyzed with FTIR to ensure that all the cutting fluid and other contaminants were removed from the MR. The study showed that washing with soap and water was more effective in removing the lubricant than using acetone, which was a better environmental choice as well.

The following variants were then compounded and plates were manufactured to conduct the study: 100% unwashed MR, 100% washed MR, 10% unwashed MR + 90% Virgin LFT, and 10% washed MR + virgin LFT. The acetone washed and soap washed were compounded as well. This resulted in a total of seven variants.

The results, although preliminary and not conclusive, show some very interesting trends. For the MR only, the washing with either acetone or soap and water had little effect on the mechanical properties. There were variations, but on average they all performed within the margin of error and about 40% lower in tensile strength and about 22% lower in tensile modulus. The results were expected, as the fiber length in the MR was significantly lower and acted more as a filler and hence the lower tensile strength and modulus.

When the MR was compounded at 10% with 90% virgin LFT, the results were a little more complicated. The unwashed and soap variants performed 17% lower compared to the virgin LFT in tensile strength and about 10% lower in tensile modulus. However, the acetone-washed MR performed equivalent or marginally better in both tensile strength and tensile modulus. It is assumed that the acetone provides a better interface between the virgin LFT and MR and also helps distribute the fiber more evenly. Further studies are being conducted to fully understand the effect.

The overall conclusion shows that the MR can be reintroduced at 10% without lowering the mechanical properties. The MR can be used on its own as a lower grade material, without any washing and/or treatment. Engineers will need to accommodate this in the design of the component.

CONTINUOUS REINFORCED TAPE TRIM-OFF TO LFT

In this study, trim-offs were received from a trailer manufacturer that used continuous GF polypropylene panels for the sides of trailer bodies. The fiber content in the panels was about 70 wt%, the panels trim-offs were received with a very thin coating of paint. The panels were shredded and the material collected for reprocessing. A section of the panel before and after shredding is shown in Figure 12.3.

FIGURE 12.3 Panel trim-off, as received from trailer manufacturer (below) and after shredding (top left) and pellets used to blend down (top right).

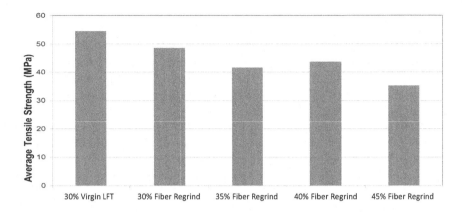

FIGURE 12.4 Tensile strength as a function of GF content in recycled PP/GF.

The regrind material was blended with neat PP at various concentrations to achieve a neat GF content of 30%, 35%, 40%, and 45% by weight. The material was blended in a low-shear plasticator and compression-molded to form plates. The plates were then sectioned and tensile tests were performed. Blending the recycled material with neat PP resulting in 30% GF content yielded the best results when compared to virgin LFT with 30% GF. The overall results are shown in Figure 12.4. There was an 11% drop in tensile strength compared to the virgin material. Considering that the panels were not cleaned and did not have the paint stripped off, etc., this is an extremely positive result. Engineers will be able to design accordingly for the reuse of this material.

CONCLUSIONS AND FUTURE PERSPECTIVES

Composites recycling has become more urgent than ever due to the tremendous amount of composite structures that are reaching their EOL and their adverse economic and environmental impacts if landfilled. However, there exist several factors, including the heterogeneous structure of the composite, the great bonding between the matrix and the fiber, the stable cross-linked chemical structure of the thermoset polymer, as well as the low cost of landfilling, limited markets for recycled composites, and less restrictive environmental legislation that have posed great challenges in the composites recycling. Although scientists and engineers have come up with some great ideas toward more cost-effective and less environmentally impactful recycling of the composite, more work needs to be done collaboratively at all levels, from material design to standardization and legislation, etc.

Applications for the use of recycled composites still pose one of the biggest challenges to the composites recycling industry. Potential new users have a major concern regarding the consistency of supply and the quality of the material received. A robust program driven by the industry itself will alleviate these and other concerns, which will lead to higher potential for reuse of the recycled material. Engineers need to be creative and look for applications that will promote the use of recycled composite materials. After all, composite materials are relatively expensive when compared to their monolithic counterparts and provide superior properties and lightweight.

A further consideration should be a design paradigm change for composite manufacturers. The first part will be to consider more standardized formulations, to make recycling easier. It is well known that manufacturers closely guard their formulations, as it gives them an advantage compared to their competitors. However, this leads to millions of different formulations and compounds the problem of recycling. Standardization will also provide recyclers with a critical mass of material to successfully consider niche applications. A vision for the growth of the entire industry should be embraced, which will give manufacturers a slice of a much larger pie.

Engineers and material scientists need to also consider recycling at the beginning of the process, rather than at the end. A successful plan should be implemented at the concept phase of the design process. This will lead to successful protocols being developed at the inception, rather than only considering the problem at the EOL of a component. Engineers should also embrace the use of recycled materials in the design of new components and products to further the use of recycled materials and close the loop with regard to sustainability.

REFERENCES

1. M. Ewall. "Primer on landfill gas as 'green' energy". Energy Justice Network. Nov. 2007. [Online] http://www.energyjustice.net/lfg/. Accessed in Feb 2021.
2. G. Gardiner. "The making of glass fiber". Composites Technology, April 2009 [Online] http://www.compositesworld.com/articles/the-making-of-glass-fiber. Accessed in Feb 2021.
3. Markets and Markets. "Glass Fiber Reinforced Plastic (GFRP) Composites Market worth $45.12 Billion by 2019". Markets and Markets. March 2014. [Online] http://www.marketsandmarkets.com/PressReleases/glass-fiberreinforced-plastic-composites.asp. Accessed in Feb 2021.
4. V. Mom. "Landfill statistics in the U.S... staggering". Think Outside the Bin. Nov. 2010. [Online] http://thinkoutsidethebin.com/2010/11/30/landfill-statisticsin-the-u-s-staggering/. Accessed in Feb 2021.
5. L. Landes, "LANDFILLS: Hazardous to the environment". Zero Waste America. Lynn Landes. 14 March 2010. Philadelphia, PA. [Online] http://www.zerowasteamerica.org/Landfills.htm. Accessed in Feb 2021.
6. Propex.com. "Landfills and hazardous waste sites". Propex.com. 2005. [Online] http://www.propex.com/C_f_env_landfills.htm. Accessed in Feb 2021.
7. Waste Management, Inc. "Tipping fee". [Online] https://www.wm.com/glossary.jsp?b=R&e=U.
8. Environmental Research & Education Foundation. "Analysis of MSW landfill tipping fees". April 2018. [Online] https://erefdn.org/product/analysis-mswlandfill-tipping-fees-2/. Accessed in Feb 2021.
9. K. Peattie. "Towards sustainability: The third age of green marketing". *The Marketing Review*, Vol. 2, No. 2, pp 129–146, 2001.
10. K. N. Hardin, S. Pillay. "A holistic approach to composite recycling". *Poster Presented at the Composites Advanced Materials Expo Conference*, Orlando, FL, Sept. 2014, 135.
11. B. Moffit. "Why recycle composites?" *Green Composites Workshop*, Chicago, IL, Jun 20–21 2012.

12. PR Web. April 2011. "Global epoxy resins market to reach 3.03 million tons by 2017, according to a new report by global industry analysts, Inc". PR Web. [Online] http://www.prweb.com/releases/epoxy_resins/paints_coatings_laminates/prweb83   43600. htm. Accessed in Feb 2021.

13. J. Li, P. L. Xu, Y. K. Zhu, J. P. Ding, L. X. Xue, Y. Z. Wang, "A promising strategy for chemical recycling of carbon fiber/thermoset composites: Selfaccelerating decomposition in a mild oxidative system". *Green Chemistry*, Vol. 14, No. 12, pp 3260–3263, Sept. 2012.

14. Y. Liu, J. Liu, Z. Jiang, T. Tang, "Chemical recycling of carbon fibre reinforced epoxy resin composites in subcritical water: Synergistic effect of phenol and KOH on the decomposition efficiency". *Polymer Degradation and Stability*, Vol. 97, No. 3, pp 214–220, Mar. 2012.

15. C. Morin, A. Loppinet-Serani, F. Cansell, C. Aymonier, "Near- and supercritical solvolysis of carbon fibre reinforced polymers (CFRPs) for recycling carbon fibers as a valuable resource: State of the art". *Journal of Supercritical Fluids*, Vol. 66, pp 232–240, Jun. 2012.

16. P. Xu, J. Li, J.P. Ding, "Chemical recycling of carbon fibre/epoxy composites in a mixed solution of peroxide hydrogen and N, N-dimethylformamide". *Composites Science and Technology*, Vol. 82, pp 54–59, Jun. 2013.

17. S. J. Pickering, R. M. Kelly, J. R. Kennerley, C. D. Rudd, N. J. Fenwick,. "A fluidised bed process for the recovery of glass fibres from scrap thermoset composites". *Composites Science and Technology*, Vol. 60, No. 4, pp 509–523, Mar. 2000.

18. J. Stike. *Reclaiming & Re-engineering Carbon Fiber Composites*. Green Composites Workshop, Chicago, IL, 2012.

19. J. R. Correia, N. M. Almeida, J. R. Figueira. "Recycling of FRP composites: reusing fine GFRP waste in concrete mixtures". *Journal of Cleaner Production*, Vol. 19, No. 15, pp 1745–1753, Oct. 2011. 137

20. A. Jacob. "Recycling of glass thermoset composites via cement kilns compliant with EU legislation". Materials Today, Feb. 2013, [Online] http://www.materialstoday.com/composite-industry/news/recycling-of-glassthermoset-composites-via/. Accessed in Feb 2021.

21. G. Nilakantan, S. Nutt. "Reuse and upcycling of aerospace prepreg scrap and waste." *Reinforced Plastics*, Vol. 59, No. 1, pp 44–51, Jan. 2015.

22. S. J. Pastine. *Sustainability by Design: Introducing Recyclable Epoxy*. Green Composites Workshop, Chicago, IL, June 2012.

23. R. J. Morgan, C. M. Walkup. "Epoxy matrices for filament-wound carbon fiber composites". *Michigan Molecular Institute. Journal of Applied Polymer Science*, Vol. 34, No. 1, pp 37–46, Jul. 1987.

24. J. E. White, H. C. Silvis, M. S. Winkler, T. W. Glass, D. E. Kirkpatrick. "Poly(hydroxyaminoethers): A new family of epoxy-based thermoplastics". *Journal of Advanced Materials*, Vol. 12, No. 23, pp 1791–1800, Dec. 2000.

13 Using Artificial Intelligence
A Paradigm Shift in Data Management

Prateep Guha and Prabir Aditya

CONTENTS

INTRODUCTION

The idea of *artificial intelligence (AI)* is not new. It has existed in many myths across time. As early as the 1960s, machines were made to simulate human behavior. The knowledge of experts was stored in computers in the form of sequenced questions and answers. However, these expert systems had limited growth in practice.

It is only recently that the ideas of simulating intelligence have been formalized with real intent for practical use. In the 21st century, AI has witnessed significant progress in applications. What was primarily a number of academic concepts and the subject for research projects began to see rapid development in practice in markets?

The three key factors which helped the growth of AI in everyday life and in businesses were:

DOI: 10.1201/9781003161738-13

- Rapid growth of *digital data*
- The fast-paced growth and the speed of the *Internet*
- Tremendous growth in the computing power

RAPID GROWTH OF DIGITAL DATA

For machines to become intelligent, they must learn. To learn, machines need data. In our time, immense volumes of data are being generated daily for machines to train on.

Petter Bae Brandtzaeg of the *SINTEF ICT* had once said that "A full 90% of all the data in the world has been generated over the last two years" [1].

This statement is still being used, after more than 7 years. For it to still be correct today, it would mean that the aggregate data volume in the world rises tenfold every 2 years!

The precise numbers might not be tenfold, or 90%—but there is no question that the amount of data is increasing exponentially. The point here is that there has been a massive explosion in data volumes in recent years.

Much of rapid data expansion comes from machines monitoring or measuring activity of humans or other machines.

We can look at a few numbers to get an idea of the sheer amount of data increase in any given unit of time. In 2020, the following happened every minute:

- Twitter gained 319 new users
- YouTube users uploaded 500 hours of video
- Amazon shipped 6,659 packages
- LinkedIn users applied for 69,444 jobs
- Facebook users loaded 147,000 photos and 150,000 messages
- Zoom hosted 208,333 participants in meetings
- Consumers spent $1,000,000 online
- WhatsApp users shared 41,666,667 messages [2]

The unbelievable growth rate of digital data meant that there was a really strong need for developing systems that processed the data. Many of these systems are what is today commonly called artificial intelligence, or AI.

ADVANCES IN NETWORK TECHNOLOGY AND THE INTERNET

The advances in network and its application in Internet technology have significantly transformed human life.

The inventor Nikola Tesla once predicted an interconnected world.

When wireless is perfectly applied, the whole earth will be converted into a huge brain [...] we shall be able to communicate with one another instantly, irrespective of distance. [3]

Today, we see Tesla's vision all around us—a world of constantly connected computers. Indeed, even the number of devices on this vast Internet is rapidly increasing. Continual advances in networking technology, along with a demand from organizations to reap the fruits of information, give us power. It is the power to extract data from diverse systems, use source programs developed by researchers, and apply our own business knowledge to develop effective AI solutions.

GROWTH IN COMPUTING POWER AND ADVANCES IN COMPUTING METHODOLOGY

AI applications are highly *computationally intensive*. Most computers throughout history were not nearly powerful enough to run AI systems. This is one of the major reasons why AI applications did not develop as rapidly in the past as it develops today.

In 1965, Gordon Moore said that the number of transistors in a given *circuit space* doubled every year. This doubling is commonly called Moore's law in the industry. The "law" held true for some time. Then, the doubling rate slowed down to every 2 years. It remained there well in the 21st century [4]. Now, there is a reason to believe that this growth rate will slow down or be halted by physical limitations. Moore's law might be on the way out. Advances in processor technology, the use of specialized processors called *GPUs*, and the use of *parallel processing* have resulted in immensely large growths in computing power—this has made AI applications feasible today.

The **exponential growth of data**, the **developments in network technology**, and the **advances in computing capabilities** are three big factors driving the adoption of AI systems across all walks of life in today's world.

Business organizations, small and large, are looking to customer transactional and behavioral data to **decipher customer preferences, understand trends and project how they will act in the short and medium term**. The *Internet of Things* has made available a wealth of previously unknown and unused information which increases the range and depth of our data literacy which can be used for organizational benefits.

In short, AI applications have changed and continue to change our lives and businesses every day.

The next part of the chapter will be talking about the building blocks of AI with the use of suitable case studies. It will discuss the various AI techniques commonly adopted by businesses. The techniques used in the examples shown here can easily extend itself to many functions across any organization.

THE BUILDING BLOCKS OF AI SOLUTIONS

Creating and applying AI solutions to everyday life and in businesses requires a multidisciplinary approach. When organizations are able to create multidisciplinary teams, they are able to harness all the data and create AI solutions which improve productivity and hence profits.

The three essential building blocks of creating AI-driven solutions are:

- Data and the capability to manage the same

- Business knowledge and ability to innovate
- Knowledge of the AI and *Machine Learning* techniques.

Data and the Ability to Manage it

The whole concept of AI is based on computers gaining knowledge through analyzing data. Availability of electronic data is growing at an unbelievable pace, and it is these volumes of information that have allowed AI solutions to give accurate results in the past few years.

Throughout this chapter, we will look at some case studies that help illustrate the marvelous things we can do with the large volumes of data we have today, and the incredible tools we have to process that data.

Use Case 1: E-Commerce Retailing

The first case study we will look at is how *e-commerce* shop owners are dealing with faceless customers. It is also a good illustration of how complex data is captured and managed.

The problem here is understanding the buying behavior of faceless e-commerce customers for small stores.

Are there techniques available to classify the prospects and customers into various categories based on their behavior when they visit the site and navigate across the pages and on a page? Can their behavior be used to establish their propensity to buy and based on their propensity score can actions be prompted to ensure the prospects move toward a purchase?

What is the benefit of implementing a proper model for this problem? The answer is that it would help increase sales volume for e-commerce stores.

Prospect/customer wise visit to the e-commerce site and their behavior while navigating through the various pages and on individual pages. The complete data for the prospect/customers' second by second mouse movement was provided.

Every day more and more new e-commerce shops are opening across the globe. E-commerce business has seen unprecedented growth in 2020.

"Despite a challenging year for retail in 2020, we estimate that worldwide retail e-commerce sales posted a 27.6% growth rate for the year, with sales reaching well over $4 trillion." [5]

Competition among the e-commerce shop owners is becoming extremely stiff each day. E-commerce stores are selling their product keeping their margins thin to keep up in the competition. With such low margins, the key to stay afloat in this business is to increase sales volume.

E-commerce shoppers are the shoppers without a face. The personality of an online shopper is starkly different when it is compared with a person physically visiting stores, picking up items, and lining up at the check stand at a physical store. Online shoppers are more merciless, leaving one e-commerce shop to another if one item is faulty.

An e-commerce shop owner's success depends on understanding the life and patterns of those faceless customers. E-commerce shop owners haven't ever seen their customers, but they have all the data to understand the customers' lives and buying patterns.

The use of AI and ML is one of the most adapted tools and techniques used in dealing with these data. Today's major e-commerce platforms are bringing new AI tools every day for shop owners to understand customers.

The use case detailed here addresses some of these challenges – how with the help of AI one can convert challenges into opportunities.

Shop owners need to understand the behavior of customers by evaluating their stickiness. Measures to aid e-commerce shop owners introduced include:

Customer Lifetime Value (CLTV): Simply speaking it represents the amount of revenue a customer will give the company over the period of time.

Churn Rate is another important measure to assess as a prelude to calculate CLTV. Churn Rate is the rate at which customers stop doing business with a company.

CLTV drives targeted **Campaign Marketing** each one is uniquely crafted for each customer. Campaign Marketing is derived from the broken-down picture of CLTV. It takes into consideration of Average Order Value, Repeat Rate, Purchase Frequency, of course.

Churn Rate. Clustering techniques are used taking all these into factors plus **RFM analysis** (Recency, Frequency, Monetary). Here customers are divided into four groups for each of the three parameters depending on which quarter of ones' customer base the individual case falls. RFM scores, therefore, range from 111 to 444 with 444 being the best. **Campaign Clustering** is done in four groups— Champion Customer, Loyal Customer, New Customer, and Lost Customer. Most of the campaigns are designed around this principle.

Cohort Analysis is another tool to measure the retention of customer groups. Above all, Pareto Principle is still the golden scale to identify those 20% customers that give 80% of total revenue.

Another key dimension of e-commerce business is to build a product mix that can vary widely between two customers. **Market Basket Analysis** is another very important Machine Learning (ML) tool. On the one hand, shop owners need to understand their customers; on the other hand, they need to understand what product their customers buy and in what combination. Market Basket Analysis uncovers associations between items together with frequency of transactions.

In one case, a shop owner was planning to get rid of some items which were slow moving and did not offer good margins. On examination of baskets and customers, it was revealed that the items were bought by the most profitable customers, who contributed the maximum to the shop owner's revenue; getting rid of them might turn the customers to other vendors.

Now, a little more on understanding data.

When we speak of the data required to build AI solutions for an organization, it helps understand the types of data available to solution builders.

Structured vs. Unstructured Data

Data can be **structured** or **unstructured**. Structured data exists in forms easily readable and interpretable by computers. Typically, structured stays in tables (with columns and rows) and allows users to filter and query the same for day-to-day use. Business transaction data in ERP is structured data.

Structured Data

Some data is stored in database tables, data warehouses, indexed files, etc., and their layouts are available in databases or application programs. This data is structured and easy for programs to analyze and use.

Business managers are very familiar with using Word, Excel, and PowerPoint. Data sitting on Excel files can be loosely termed to be structured. Here, the user is able to sort, filter, and extract data based on specific data conditions. For example, if manufacturing batch-wise quality data is kept in Excel files, one is able to easily count which batches produced good quality out and which didn't. This is possible since the data is structured.

However, if an organization has all its offers to customers and prospects sitting on Word files, it would be very difficult to find a count of the different payment terms applied or the position the organization has taken on the various liability clauses. Data in Word files is unstructured and it is very difficult for the computer to read the documents and arrive at those counts.

Unstructured Data

Apart from computer transactions, organizations today have a lot of information stored on computers. Emails, market surveys, legal documents, contracts with vendors and consultants, strategy documents, and code of conduct all form a wealth of information which is in text format and require programs to use routines which understand "natural language," to create data that can be used for analytical applications.

Pictures are another source of information, and image recognition and processing are another requirement for today's AI systems.

Let us try to understand the problems traditional programs have with unstructured data.

- There is no defined format and as a result, the meaning and relevance of any piece of information contained have to be understood by analyzing the text prior to and after it.

- Unstructured data can contain implicit as well as duplicate information, which must be understood by analytics before any actionable information can be created.
- Numbers may be represented numerically or alphabetically (Five and 5) or by a mix.
- Identifying words which are proper nouns is often problematic.
- The context often defines the meaning of words: Two plus four is six refers to numbers, he was wearing plus fours, refers to a type of dress.

The complaint mails from customers of an organization are an important source of information on product quality. However, traditional programs will not be able to use them to detect types and volumes of defects.

Or take, for example, the problem of processing written contracts in organizations. Traditional programs have trouble with these because the data is unstructured. But natural language processing systems can structure the unstructured documents so that it can be read, interpreted, and processed by other programs—these engines can be used to pull relevant information out of messy, chaotic, convoluted, and unstructured Word documents.

The velocity and volume of data in recent years has been enormous, mainly due to two sources—(1) social media and (2) Internet of Things (IoT), where sensors capture data from devices every second and constantly send this data to data storage structures.

Use Case 2: Fitness Trackers

The next case showcases an AI solution to data acquired from IoT devices, specifically wearable devices.

The problem here is tracking of performance for trainees inducted for defense and the police academy. The objective is to monitor how each member is performing in terms of physical fitness and punctuality to attend scheduled programs.

The benefit—data-driven performance analysis to assist evaluators, coaches, and mentors for accurate evaluation.

The data we are working with includes wearable device stream different parameters, such as heartbeats, walking running, jogging speed, sleep patterns, calories intake, and calories burnt. Seventeen data parameters are used to perform the analysis.

The academy follows a few strict guidelines to attain standardized physical fitness levels, punctuality, and discipline.

Data is transmitted from wearable devices in different modes of communication channels, such as *WiFi, GPS, Bluetooth*, and *Mobile Network*. Data is transmitted as live stream or batch push mode as and when required.

Each member's basic vitals are captured as master data, such as age, gender, height, chest, waist, body ratio, etc.

The health index is also an important aspect of measurement. Heartbeat counts during different outdoor physical activities are captured from the devices and automatically gets translated into fitness indexes.

The diet plan is fed to the system. Calories consumed per day are calculated from the diet sources. Calorie burnt information collected from devices is compared. Charts are prepared and correlated between food intakes and physical fitness and weight gain/loss.

Multiple AI models are used to create multiple success factors. Models are related to correlation and regression, clustering, time series analysis, etc., are used.

So, given the sheer amounts of data available to use in such optimization problems, a natural question is—how is all this vast data stored?

Data Storage and Manipulation using Big Data Technology

Traditionally data was stored and processed from large servers. With the humongous growth in data volumes and types of data, the need for analyzing the data to understand data patterns, associations, clustering, and trends became a possibility. The growth and the speed with which the data had grown made it very difficult for traditional data-processing techniques to process, compute, and extract information meaningfully.

Thus the emergence of Big Data Technology enabled data of all types to be stored in a distributed way on multiple servers and allowed the users to use the power of all the storage and the memory and the processing power of all the servers simultaneously for various business applications at speeds which one never dreamt of earlier.

We will next look at the second building block of implementing successful AI solutions—business knowledge and innovative ability.

Business knowledge and ability to innovate

Expertise in specific subjects, or domain-specific knowledge, is essential in building AI applications. We need domain experts in AI that know the nuances of individual business functions.

AI solution builders use the services of experts in different domains to build their business solutions. Customization to the specific needs of individual organizations requires inputs from the process owners.

The role of the subject matter expert is to:
• Precisely state the problem
• Identify the intricacies and dependencies
• Identify the data sources available that are required to build the solution
• Comment on the applicability and feasibility of proposed solutions
• Break down complex business problems into smaller, simpler components
• Give the final thumbs-up when the solution gives expected results after it is applied to past data (training set).

Use Case 3: Optimization of Fuel Consumption in Turbines

The next case study illustrates the essentiality of subject matter experts in constructing an AI solution. In this case, the goal was to optimize fuel consumption at a petroleum refinery, thereby reducing costs and increasing profitability.

A petroleum refinery plant used a number of steam turbines fed by boilers for each of the turbines and a series of gas turbines for its captive power needs for the entire plant. Monthly and weekly demand for the power for the entire plant was computed from the target figures plant management. Based on the power requirement, the plan had to operate its steam and gas turbines to feed the demand. The management had a feeling that the fuel consumption for the boilers and gas turbines was not optimized and there existed an opportunity for cost savings using mathematical models.

The huge benefit from implementing AI models using the data from operations of boilers and turbines is significantly reduced costs due to optimized fuel consumption.

Operating data for all the boilers such fuel flow, water flow rates, steam temperature, heat generated, etc., and for turbines including fuel flow, gas supplied for the combustion, power generated, and exhaust gas composition, etc. This data was provided for a period of 1 year at an interval of 1 minute. These time spans give an intuitive idea of just how much data we are working with.

AI models were created for each of the boilers and the turbines based on the data provided. Multiple supervised models were created using different AI techniques for each of the boilers and the turbines with part of the data provided. The balance data was used to test the models for their efficacy. The model suggested the optimal use of fuel and water for the boilers and for the turbines it recommended an optimized fuel flow rate and input gas. Significant fuel savings were achieved for the boilers as well as the gas turbines.

Since structured data were made available in a time series for the boilers and the turbines, it was possible to easily clean and organize the data and feed the same into the AI methods to build the models.

Once the model for the individual boilers and turbines were ready, the second part of the problem, which was meeting the daily power requirement for the plant from the turbines, had to be cracked. The turbines, both gas and steam ones, had different capacities. Again a model was used to recommend what combination of the turbines should be used to meet the daily/ weekly and monthly power requirements.

The ROI for creating the model was achieved within a period of 5 months of the model creation.

We will now look at the final building block of implementing successful AI solutions—the tools themselves.

AI Branches, Methodology, and Tools

Machine learning is the field of AI that provides systems the ability to learn and improve from experience without being explicitly programmed.

Machine learning can be segmented into three divisions—supervised learning, unsupervised learning, and reinforcement learning.

As the name suggests, **supervised learning** refers to training a computer program on what conclusions to arrive based on a representative sample set of data, called the training set. The two major methods of supervised learning are *classification* and **regression**.

Unsupervised learning refers to methods where the computer has to go through data without supervision and find patterns previously unknown. Important methodologies in this segment are *association, clustering, dimensionality reduction,* and *density estimation.*

In **reinforcement learning**, algorithms need to work out a number of decisions in sequence, each with its merit or demerit points with a view to maximize, through trial and error, the cumulative reward.

Natural language processing deals with the interfacing of people and machines using human languages (which in this domain are called natural languages). The overarching aim of NLP is to make artificially intelligent systems that can totally comprehend natural languages. NLP lends itself naturally to useful tools such as text summarization and keyphrase extraction. A commonly talked about structure in NLP is the NLP toolkit. NLP toolkits, among other things, provide a corpus of natural language words and linguistic structures, along with their associations and semantics—what they mean, what they symbolize, relationships between them, etc.

Use Case 4: Chatbot for Insurance Call Center

The following is a case study that illustrates how natural language processing can be leveraged in the insurance industry.

Insurance companies maintain call centers. At these call centers, call agents attend to varied customer and prospect questions. Any delays or incomplete responses lead to customer or prospect dissatisfaction and result in revenue loss from customer attrition.

Fixing delays and incomplete responses means improved customer relationship management and hence better customer retention.

The data we work with here includes recording of past customer calls to call center agents along with complete details of all customer details, their insurance policies, and all past transactions.

The solution involved using an intelligent *Chatbot*. Customers were allowed to place their questions on the bot and have immediate automated responses from the bot.

Creation of the intelligent chatbot involved a detailed study of past recording of agents responding to customer queries, policy details, and customer data. Based on the queries posted by the customer, the bot was able to respond accurately to customers—more promptly and accurately than an agent with a faster turnaround time than individual calls. Any query which the bot was not able to respond to was transferred to an agent.

Benefits derived were manifold:
1. Reduction in response time to customer queries
2. Reduced staff requirement at help desk

3. Record of all queries leading to efficient update of FAQs and responses
4. Complex queries requiring human interaction handled faster

All this helped in reducing costs and customer attrition.

The AI-assisted chatbots comprise a series of AI models blended with natural language processing engines semantically enabled. These consist of a judicious combination of rule-based logics and hierarchical entity flow built on a foundation of several thousands of real-life situational data points.

Another AI technique is *Deep Learning*. Deep learning is a powerful methodology that aims to mimic computations of the human brain. Deep learning uses structures called *neural networks*, a complex set of interconnected nodes that feed into each other to process data. Neural networks have many variations and operate on both structured and unstructured data. The uses of neural networks span many domains—a significant one is their implementation in computer vision.

The next three case studies show the varied implementations of neural networks.

Use Case 5: Consistency of Product Quality in Composites

The first case tackles the problem of measuring product quality and consistency of composite materials.

Composite materials are produced using more than one type of fiber. During production, uniform distribution of the different fibers may vary and result in impacting the physical strength of the composite.

Early identification of the uniformity and segregating the ones where mixing is not uniform would significantly help in improving product quality, thereby reducing costs.

So, how does one measure the uniformity of the different fibers in the composite?

Multiple cross-sectional images were provided for different types of composites.

This is a typical case of using unstructured data being structured for problem-solving using AI techniques.

Microscope images of the cross-section of the composites are taken. For a reasonable sample of the microscopic images, the individual fibers are identified.

The annotation process involved the fibers being identified on the microscope sectional images on the samples.

The technique used for building the model is called a *Convolutional Neural Network*. The model is trained using a large sample of the annotated images. The model learns to identify the different fibers, their texture, color, typical size, and shape. Algorithms were created to build a measurement metric for the comingling of the different fibers.

Once the solution was created, the model was capable of numerically expressing the consistency of the mix for any cross-sectional microscopic image for the composite for which the model was created. For any metric for the composite which was below the acceptable level, problem identification was early and manufacturers were able to tie the same to the manufacturing process for the composite to eliminate the problem. A quantified measure of consistency also allowed manufacturers to move away from subjective judgments of the product.

Use Case 6: Neural Networks for Transportation and Logistics
Neural networks have also been effectively used to tackle problems in the logistics industry.

Planning pickups and drops of multiple shipments in different locations is complex and the chances are that truck capacities and driver availability are underutilized.

Fixing this optimization problem would mean cost-saving and timely delivery of shipments, leading to customer satisfaction.

The data available here is manifold—including roadmaps, truck and driver availability, truck capacities by volume and weight, dimensions and weight of shipments required to be transported, and special considerations (such as refrigeration requirements).

Transportation and logistics planning was one of the first areas which AI has sought to address.

Effective utilization of fleet by intelligently combining multiple drops in single trips and consequent route planning is now being achieved by the use of neural networks to address complex problems with several variables (multiple pickup and drop points, multiple possible routes, truck availabilities and capacities, products which can or cannot be delivered together, driver availability, refrigeration requirements, etc.). These have yielded solutions which are being put to use by many transport organizations.

AI-based transportation solutions today look to reduce chances of trucks traveling empty, thereby improving productivity and reducing costs.

Use Case 7: Neural Networks for Structural Health
The final case using neural networks involves monitoring structural health of large steel structures.

The problem is the structural deterioration due to aging, corrosion, fatigue of materials, extreme environmental loads, and unexpected impacts on metal structures.

The solution can be achieved by the use of unmanned aerial vehicle-based remote sensing and digital image processing.

Noninvasive diagnostics of steel structures using unmanned aerial vehicles (UAVs) are now becoming popular. The present case study describes work that consists of image acquisition, preprocessing, segmentation, extraction of features, and quality assessment. Several factors that may affect the acquisition process have been carefully analyzed. Adequately selecting the focal length of the camera, adequate close-up of details, ensuring the orthogonality of the optical axis to the examined object, and eliminating unfavorable external factors such as atmospheric conditions are considered for analysis. AI models were used for detecting visible signs of structural fatigue and implementing machine learning.

The automotive industry has also gone in for AI in a big way. Let us look at what is happening there.

AI AND THE AUTOMOTIVE INDUSTRY

"Global sales of automobiles are forecast to fall to just under 64 million units in 2020, down from a peak of almost 80 million units in 2017" [6].

There are many challenges facing the automobile industry today, and automakers are looking for AI solutions to remain competitive. Well-known brands stand to lose out to automakers who have implemented the latest AI technologies. These technologies could assist drivers and provide increased safety features, or improve the ride experience of drivers and passengers.

Driverless cars have been built and tested on the field, but there are no plans for mass production at the moment. However, AI-enabled ADAS (advanced driver-assistance systems) are being introduced by all major car manufacturers to enhance and improve driving and ensure vehicle safety.

Features that are being introduced include automatic parking systems (which take full control of parking functions until the vehicle has safely checked and negotiated other cars and obstacles and positioned itself into an empty parking spot), traffic sign recognition, rain sensors, checking tire pressure, automatic braking in emergencies, elimination of blind spots, alerts when pedestrians or obstructions are spotted, lane crossing alerts, collision warning, automatic slowing when taking corners and coming up on toll booths, detecting speed limit signs, and automatically adjusting to the limits.

Solutions to improve passenger experience include voice-activated virtual assistants which control the temperature, switch on lights, play music, and entertain children.

AI solutions also exist for automobile manufacturing (defect detection through computer vision, modeling, and prototyping), supply chain (demand prediction and automated order processing on multiple component manufacturers), and connectivity (sensors providing information to manufacturers on performance and alerts on possible component failures).

AI PROJECT STAGES

An AI project could get initiated in many ways. Often, it gets initiated from the business, but depending on the data and AI strategy maturity of an organization it could also get initiated by the organization's management services and IT departments. The flow chart depicts the various phases in any AI project.

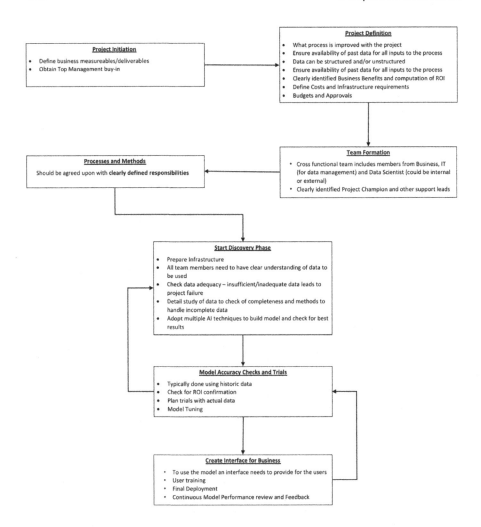

The phases in any AI project are a function of the organization maturity in its data and AI strategy. Mature organizations would have clearly established processes.

The success of the project is a function of numerous factors—Top Management commitment, the vision of the business team, easy availability of the data and finally, skilled AI professionals.

CONCLUSION

To benefit from AI, enterprises need to have firm strategies for all the three building blocks which should in turn be aligned to the corporate strategy. The data strategy would help enterprises to have all internal and external data, structured and unstructured made available to the business and AI practitioners. The Business Strategy would enable the enterprise to look at the availability and innovate new ways of working. Coupling the above AI strategy would be able to create the valuable AI solutions for the enterprises' *digital transformation* journey.

As shown in the various use cases in this chapter, AI aids enterprises in forecasting, identifying patterns in data both structured and unstructured, finds associations, identifies the most important factors which control an event, it helps in finding clusters of data, it helps in recognition of objects and separates out the dissimilar ones, and in many more ways. This chapter did not go into the complex mathematics and statistics behind the various techniques, which are best left to the technical AI professionals. Instead, it attempts to lay out the building blocks of creating an AI solution for an enterprise.

The AI case studies that have been used in this chapter would enable business managers to draw parallels in what they do and identify use cases which could be possible candidates for AI implementation. The chapter should help non-AI professionals to build the right core team to identify business cases and discover ways to reduce costs or improve productivity and build on profitability.

In today's world, "Digital Transformation" is a buzzword which organizations are grappling to adopt. Our experience with enterprises across the globe tells us that implementation of Digital Transformation is happening in many ways across different organizations. The most ambitious organizations could take the approach of building a huge data repository which captures data and information, structured or unstructured, across the length and breadth of the organization. This would be the outcome of the enterprise data strategy. Functional Managers are thus enabled to innovate how they could use the data for business benefits, which would include the adoption of AI techniques for creating insights from the data that is available in the central repository. Conservative organizations take a modest approach and approach and look at individual business use cases and in the absence of a data strategy, work out a plan to have the data organized and then adopt the AI technique to create a solution.

GLOSSARY OF TERMS

Artificial Intelligence: Artificial intelligence refers to building computer programs to simulate human intelligence including reasoning and learning from events like humans.

Association: Association rule learning is a method to discover and establish meaningful relations between variables in large data sets.

Big Data: Very large data sets which need computationally intensive analysis to reveal patterns, trends, and associations.

Bluetooth: A wireless technology methodology used for exchanging data between fixed and mobile devices over short distances using radio waves.

Chatbot: An artificial intelligence application that mimics a human to conduct an online conversation with a user in the user's natural language.

Circuit Space: The area enclosed by a circuit.

Classification: The categorization of observations or entities into sets (categories, subcategories) based on their characteristics.

Clustering: Clustering is a method of grouping the variables in large data sets based on their similarity on specific aspects.

Computationally Intensive: Calculations involving very large numbers (in size or count) requiring considerable CPU time to process can be called computationally intensive.

Convolutional Neural Network: A Convolutional Neural Network is a special type of deep neural network used mainly to classify images, perform object recognition, and cluster images by similarity.

Correlation: A statistical measure that gives the degree to which two variables are related linearly.

Deep Learning: A neural network with multiple layers of neurons where the neurons in each layer process input data from previous layers hierarchically and pass on more complete information to subsequent layers is known as a deep neural network and the methodology of AI which uses this is known as deep learning. A deep learning system is unsupervised and self-teaching.

Density Estimation: Statistics is the science of making estimations about a population from a "representative" sample. Probability density estimation refers to estimating the probability of your calculations being accurate.

Digital Data: Data recorded or transmitted in the form of innumerable minute discrete signals

Digital Transformation: Digital transformation is the transformation of business and organizational activities, processes, competencies, and models through an integration of digital technologies.

Dimensionality Reduction: Dimensionality reduction refers to methods to reduce the number of input variables in data sets. This is required when there are too many variables (or characteristics) in the input of which some may not have meaningful relationship with the outcome variable. An important method of dimensionality reduction is principal component analysis (PCA).

E-commerce: Business transacted online (on private networks or Internet)

GPS: Global positioning system—a system of satellites, ground stations, and receivers which are able to accurately calculate the position of a person or object.

GPU: Stands for graphics processing unit, consisting of a large number of cores, which can do multiple (thousands) operations at the same time. Applications include computer graphics, video processing, and now, artificial intelligence computing.

Internet: A global wide-area network that connects computers and other electronic devices across the world.

Internet of Things: The network of computing devices; mechanical and electronic machines; objects with sensors; implants in humans, animals, or plants, each with a unique identifier in order to share data and instructions.

Machine Learning: The field of artificial intelligence that provides systems the ability to learn and improve from experience without being explicitly programmed.

Mobile Network: Technology used to connect remote users through a network operator where the last link is wireless communication. It consists of a network of base stations, each covering a defined area (called cell) and routing communications to and from users' terminals using radio waves.

Natural Language Processing: Natural language processing or NLP is a field of artificial intelligence which enables computers to read, understand, and derive meaning from human languages.

Neural Network: A neural network is a connected set of computer algorithms that imitate the way the human brain works to recognize and understand underlying relationships in sets of data. Human brains are made up of connected networks of neurons, and ANNs (artificial neural networks) attempt to simulate them to take human-like decisions.

Parallel Processing: The technique of breaking down a computer task into several components and executing them at the same time on multiple computer processing cores.

Regression: A statistical method which determines the relationship of a dependent variable with a set of independent variables.

Reinforcement Learning: Algorithms which work out a number of decisions in sequence, each with its merit or demerit points with a view to maximize, through trial and error, the cumulative reward

SINTEF ICT: SINTEF is a multidisciplinary applied research and development organization based out of Norway. SINTEF ICT is a department specializing in information and communication technology.

Social Media: Applications that network (traditionally Internet-based) human beings to share ideas, opinions, and information such as Twitter, Facebook, WhatsApp, Instagram, and Internet blogs.

Structured Data: Data stored in a structured manner consisting, in most cases, of records (rows) with different fields (columns) such as database tables.

Supervised Learning: Training a computer program on the conclusions to arrive based on a representative sample set of data called the training set.

Unstructured Data: Data stored without defined structures, such as word documents, text files, e-mails which traditional computer programs are unable put to use in their calculations.

Unsupervised Learning: Methods where the computer program goes through data without supervision and finds patterns previously unknown.

WiFi: A technology which allows computers, smart phones, televisions, and other devices to connect to the Internet or to communicate with one another through wireless communication within a particular area.

REFERENCES

1. https://www.sciencedaily.com/releases/2013/05/130522085217.htm Accessed Feb 25, 2021.
2. https://www.domo.com/learn/data-never-sleeps-8 Accessed Feb 25, 2021.
3. https://paleofuture.gizmodo.com/nikola-teslas-incredible-predictions-for-our-connected-1661107313 Accessed Feb 25, 2021.
4. https://www.intel.com/content/www/us/en/history/museum-gordon-moore-law.html Accessed Feb 25, 2021.
5. https://www.emarketer.com/content/worldwide-ecommerce-will-approach-5-trillion-this-year Accessed Feb 25, 2021.
6. https://www.statista.com/topics/1487/automotive-industry/ Accessed March 1, 2021.

14 Innovation and Improvement from "Inside the Box"

Amnon Levav

CONTENTS

This chapter comprises five sections:

1. Introduction
2. Fixedness: what it is and how to break it
3. Examples: general and SMC
4. Three SIT Tools and two challenges
5. Conclusions

DOI: 10.1201/9781003161738-14

INTRODUCTION

In this chapter, you will encounter a methodology for innovation called SIT – Systematic Inventive Thinking™. To be more precise with our terminology, we will define what we mean by "innovation."

Working definition: *Innovate (v.)* – think and act in a different way to achieve your goals.

Let us clarify the terms that appear in our definition:

- Think and Act: innovation obviously has a cognitive component, but it must express itself through action.
- Achieve Your Goals: innovation is considered as such only if it is a means to extrinsic ends.
- Different: this a key term in our definition since not every action that promotes one's goals can justifiably be considered to be an innovation. Plenty of useful activities take place each day in the life of a company, many of them leading to significant improvements, but we would feel awkward referring to them as innovations. What, then, makes an idea "different" enough to qualify as an innovative idea? In the next section, we offer our answer to this crucial question.

This definition has evolved during 25 years of applying the SIT method, by the eponymous company. SIT as a methodology stems from a method named TRIZ, originally developed in the former USSR by Genrich Altshuler, based on the observation that seemingly unrelated inventions actually share underlying common structures or patterns. These patterns, once identified, can be utilized to shorten and facilitate the generation of novel ideas and solutions. SIT took this idea and Altshuler's initial 29 identified patterns and defined 5 basic thinking tools, with a set of principles that govern the application of these tools. Later, the SIT method was expanded to include a variety of techniques for applying the tools in group settings and eventually to support the implementation of ideas and the creation of a sustainable culture and practice in organizations.

FIXEDNESS

A key term that does not figure in TRIZ and has been introduced in SIT's approach is the concept of "Mental Fixedness." This term was initially coined by the Swiss social scientist Karl Duncker. Duncker conducted an experiment in which participants were asked to attach a candle to a wooden wall utilizing matches and thumbtacks. The group that received the matches inside the matchbox tried to either attach the candles directly using the tacks or light the candles to use their dripping wax and glue them to the wall. Meanwhile, the second group, who received the matches apart from the matchbox, used the empty matchbox as a small shelf to sustain the candle (see Figure 1 below). This led Duncker to define "Functional Fixedness" as the tendency to ascribe to objects specific functions to the exclusion of the capability to imagine them performing other functions.

Visual of Candle and Box Solution

The concept of Functional Fixedness was published in the 1940s and has since been widely studied in a variety of countries and contexts. SIT adopted the term as one of its guiding principles and identified two additional types of fixedness, structural and relational, that limit thought in different ways. Going back to our working definition of "innovation," then, we can utilize the concept of fixedness to give a more precise meaning to the notion of "thinking differently." Thinking differently means having thoughts that break one or more of one's mental fixedness. Our recommendation, therefore, to companies that wish to decide whether certain actions or results should be deemed to be innovative is to check to what extent they comply with two concrete criteria:

1. The action or result has a demonstrable impact on achieving an existing goal
2. Arriving at the action or result required breaking one or more mental fixedness

This definition is useful, for example, for conducting innovation competitions, and it adds structure and clarity to the message when organizations call on their associates to "be more innovative." But, although raising consciousness to one's fixedness is important for enabling novel thinking, it is not a sufficient condition. When asked to identify Functional Fixedness in their own organization, for example, people tend to come up with two types of instances: they either point to cases in which *others* suffer from the phenomenon, rather than themselves, or describe cases of fixedness that they themselves have had in the past but have since learned to *overcome*. Only in very rare cases is a person able to detect a live and current instance of themselves suffering from functional fixedness. This is because fixedness is part of the very structures that allow one to comprehend the world and is, therefore, rarely transparent to introspection.

Companies are aware of the difficulty in breaking out of well-worn grooves and strongly entrenched habits, of course, and they, therefore, engage in various practices to overcome this weakness. The following are some of the more common measures used by companies to foster innovation, as exemplified in the compound materials industry:

- Mergers and Acquisitions: the major players of the 1970s – The Budd Company; General Tire; Goodyear; Eagle Picher; Rockwell International; Premix International; Bailey Corporation – were all both making molded compound and using it to create molded parts for the automotive industry. During the years, the industry has seen acquisitions (Budd by ThyssenKrupp), divestments (Premix to Bailey), and investment with consolidation (Cambridge then Meridian and then the creation of CSP and its acquisition by Teijin). These activities, at times but not always driven by the need to innovate, do, indeed, create some fixedness breaking, for various reasons. Knowledge from different industries is thrown together (Teijin brings in knowledge of carbon fibers), R&D departments are challenged by new owners, companies wish to beef up their value by boasting new technologies, and evolutionary forces push out weaker competitors forcing them to find new niches. No doubt that all this moving and shaking in the past 40–50 years has driven some level of innovation in the composite arena. Nevertheless, much has stayed constant, witness, for example, the fact that even relatively diversified key players are still very strongly automotive focused, even though it is has taken a strong lead in terms of diversification to a variety of other industries.
- A phenomenon related to M&As is that key personnel and executives use a merger or acquisition as an opportunity to branch out and create small- or middle-sized businesses that allow themselves to break into new ground. In the case of composites, these companies found it easier to venture into new territories, such as marine and other nonautomotive applications. On the one hand, freedom from the constraints imposed by larger companies tends to open novel possibilities, and on the other hand, it is easier to bring composites into new markets rather than compete head-on with the giants in their traditional ones.
- Following the innovation of a client (OEM) is also a common innovation tactic, as seen, for example, in the case of Ford and then Toyota with composite truck boxes. This enables large companies to embrace innovations relatively safely, without needing to challenge their own thinking.

All of these dynamics are useful and effective but occur haphazardly and depend on external forces. What if companies could break fixedness systematically, by following well-defined processes? Thriving companies develop the ability to innovate organically, with their personnel and resources. That is when structured approaches to breaking fixedness specifically and to innovating, in general, are brought to bear.

The Thinking Tools and other elements of the SIT methodology are designed with this purpose in mind. The tools are applied within a framework that we call Function Follows Form (FFF), following the work of Ronald Finke et al.

FUNCTION FOLLOWS FORM

Back in the early 1990s, a group of psychologists led by Ronald Finke made a fascinating discovery: When it comes to creating, people are innately better at uncovering the potential benefits of a given form than creating a new form to satisfy a given need. This discovery spurred a new approach, called Function Follows Form, that encourages one to first create a virtual situation (form), and only then explore its potential benefits (function).

From penicillin to Ivory Soap and chocolate chip cookies, history is full of inventions triggered by lucky accidents. By innovating using the FFF process, you can systematically engineer "lucky accidents" in your products and critically explore how to extract real market value from these. When you start to become aware of the potential value in a product that looks a bit strange or begin to assess the possibilities in a situation that has gone wrong (without outright rejecting these or trying to fix the "problem"), then you are adopting FFF as an innovation mindset.

Conventionally, product innovation begins with consumer need identification that is then translated into functions that the product, service, or process is expected to deliver. The product's form is then designed to fit these functions. Consumers, however, struggle to articulate unmet needs and imagining a product that does not yet exist is extremely difficult. So, although the regular, i.e., Form Follows Function, approach is very useful for creating products that fit existing and articulated needs, an additional, complementary approach should be used if one looks to create innovation that exceeds expectations and surprises competitors and users.

This opposite approach is FFF. Start with a clear existing situation (e.g., a process, a product, and a strategy) and manipulate it by applying an SIT thinking tool. The result is a virtual product or your trigger for innovation. This is also called a preinventive form, as it is not yet an idea or an invention, but merely a "form."

Once the trigger is created, it should be visualized: how *would* this look? How *could* it look? The next two steps are questions. The first: What value, opportunity, or benefits could the virtual product provide (in SIT parlance, this is the Opportunity Filter)? If you cannot find any initially promising answer to this first question, drop the trigger and go on to consider another one. But if, as is usually the case, the answer is positive, continue to ask: What will it take to make the idea happen and realize the opportunities inherent in the trigger? (This we call the Viability Filter.) These questions, in addition to filtering out less-promising ideas, serve as guides for adapting the virtual product on its way to becoming a concept that can be developed. If an idea has made it this far, it means that it both has potential as an opportunity and is, at least theoretically, feasible. Not every idea is good, but any idea that makes it this far has at least some potential, because it has passed the two Preliminary Filters (Opportunity and Viability). There are two reasons why this process is called FFF. First, because the starting point is not the function, but rather the form – an existing product or system. Second, because the form is manipulated first, and only then are its potential functions considered. The result of mentally manipulating the existing product is called a Virtual Product or trigger. Visualizing the virtual product is not a simple task since it is initially not at all clear what its possible uses could be. This can very often create a sense of discomfort due to the feeling of uncertainty generated. That, however, is the very reason why FFF is so effective. Users are forced to genuinely try to figure out possible benefits – which often leads them to identify previously unidentified needs or audiences for new products.

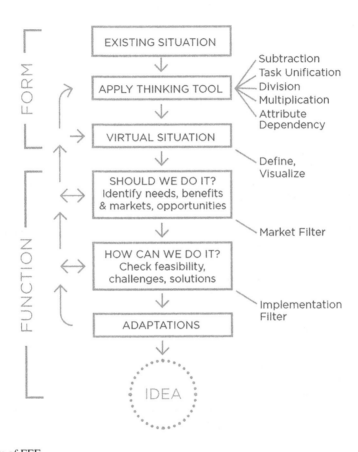

Diagram of FFF

Let us first observe an example from an unrelated field, to acquaint ourselves with the concept. Imagine a regular bicycle.

Imagine the following two triggers, as invitations to innovate on the bicycle:

A. Imagine a bicycle with no pedals
B. Imagine a bicycle with no wheels

Both sentences seem odd, and one is inclined to immediately recognize their faults as practical ideas for innovating on bikes. But, as explained above, one must resist

Visual of Bicycle.

this cognitive pull, and instead, ask oneself: what benefits or uses could be envisaged based on each of these triggers?

A. bicycle without pedals exists – it was invented and marketed in Europe as a "Balance Bike" under the name "LIKEaBIKE™," and from personal experience, I can attest that it is an extremely useful device.

Anyone who has tried to teach a child to ride a bike knows that the main challenge for the child initially is to control their balance. Many of us have run behind their child, with or without attaching a stick to the bike, sometimes for quite a long while. What did industry offer us traditionally? Training wheels. This is, generally speaking, the most common response to a bug in a product: add components to fix the problem. LIKEaBIKE™ adopted the opposite approach. By taking away the pedals, the child is obliged to propel themselves by pushing with their feet against the ground. After a while, they will start also raising their feet and cruising along with the speed created by this movement. Within weeks, they learn to control their movement and achieve perfect balance and are thus ready to upgrade to a real bicycle without passing through the training wheel phase or challenging mummy's or daddy's back. Think back to your initial reaction to the trigger. If, like most of us, your very first reaction was negative, the method is inviting you to consider your fixedness.

Trigger B – a wheel-less bike – may offer an even greater challenge to one's mental flexibility, for what is more essential to the concept of a bicycle than the fact of having wheels? But, try to follow the rules of FFF and make a sincere attempt at identifying uses or benefits. You may find yourself (re)inventing the exercise bike, which can be seen as a regular bike, from which the wheels have been taken away and a mechanism put in to stabilize the device while providing a controlled experience that simulates riding.

What we have seen with examples A and B is that what can initially seem to be a ridiculous suggestion can in fact serve as a useful trigger leading to novel and implementable ideas. These triggers were not randomly generated; they are, rather, the result of applying one of the five tools to the product – bicycle – as it existed before the invention of, respectively, training and exercise bikes. In the next sections, you can read about some of these tools, and you will surely recognize which one was used in these two cases. But before we delve into the tools, the next section invites you to challenge yourself further, by considering examples of Virtual Product or triggers that are more "painful" since they are taken from your own professional realm.

The following are several triggers for challenging your fixedness in the context of SMC – created by applying the five tools with near total ignorance of the field, just to demonstrate the possibilities and challenges of considering alternatives to the current status. Take a few minutes to review them while observing their effect on your thinking. Do you tend to reject them outright? Do you immediately line up a number of excellent reasons why each one of them cannot make sense? Can you force yourself to, at first, seriously consider each as a potentially viable option? Can you spend a couple of minutes imagining potential benefits of each of the triggers for your company? For OEMs? For you? For final users? If you manage to answer the benefit questions positively about one of the triggers, spend another few minutes answering the next question – about its viability. At this stage, you are invited to enter into a kind of negotiation. Try to find solutions that will allow you to realize the promise in the virtual situation or trigger, maybe giving up some of it, or adapting in some way.

SMC TRIGGERS

1. What if the transportation industry, for whatever reason, stopped using SMC thus forcing companies to find new markets, outside of this category? Which other industries could become major users of this type of material? Does the composite industry have to be inextricably and necessarily interrelated with the automotive industry?
2. Try to imagine an SMC product that does not include a fiber of any kind. Which kind of material – not a fiber – could play an analogous role in the process on the matrix?
3. Currently, the higher the aspect ratio of a fiber, the better, and that obviously makes much sense. Can you imagine a situation in which a lower aspect ratio becomes an advantage or an asset? Can this lead to new concepts in SMC? New offerings for the OEM or end-user?
4. Can aspect ratio change over time?
5. Can you create a composite with several levels of matrices and fibers – a triple/quadruple decker, but each level with different characteristics and therefore contributing differently to the product's features?
6. Currently, SMC is usually affected negatively by its exposure to UV or other types of light. The most one can do is to try to make the material resistant to its exposure. In contrast, could you imagine a material that actually improves with UV? Where UV activates some positive characteristic or feature of the SMC? Where the SMC adapts itself to the UV.

These triggers were created using some of the five tools, three of which are described below. After reading them, you are invited to identify which has been used to create each of the SMC triggers above. The next challenge, of course, is to try to apply one or several of them to different aspects of your business: materials, process, business model, creating your own triggers to challenge your thinking and that of your colleagues.

THREE TOOLS

Subtraction

The idea in Subtraction is to remove an essential component from a product (process or system) and then find usages for the newly created "virtual product" or trigger.

People naturally tend to add features in order to innovate, while the act of Subtraction is usually reserved for eliminating components that are responsible for undesirable characteristics. While beneficial when accomplished, the ideas created through this kind of removal process are not usually considered innovative and almost never lead to breakthroughs. In contrast, SIT's Subtraction tool helps you break your fixedness and innovate by compelling you to sacrifice a *desirable* element in your product, leading to a trigger for novel thought.

Applying Subtraction

1. List the components of the product (at least 5–8).
2. Remove a component that seems to be essential. After subtracting an essential component, one naturally tends to look for a replacement to fulfill the missing component's "dangling function." In Subtraction, we try first to subtract both the component and its function, without finding a replacement. Only after exhausting this line of thought, do we search for a replacement.
3. Visualize the resulting virtual product. When you do search for a replacement, try first to replace the eliminated component by using readily available resources – those found in the "Closed World." Try not to import external technologies or components until you have exhausted these possibilities and only then move on to explore external replacements.
4. Identify the Virtual Product's potential benefits, uses, opportunities, markets, and advantages (Benefits Filter)
5. Apply the Viability Filter (can we make it happen? how?).
6. Make necessary adaptations.

Subtraction Tips

1. Start out by subtracting a truly essential component – one that feels ridiculous to subtract.
2. Even when the virtual product sounds ridiculous (a bike without pedals or wheels), give it serious consideration. Who would have thought that a book store without a store and without books could make sense?
3. Sometimes only a part of a component is subtracted (partial subtraction). This option is often easier to digest versus a full Subtraction. Still, give yourself a chance to grapple with the tougher challenge before you soften the task and go for partial.

Multiplication

In Multiplication, you add to a product (or process or system) additional copies of a component of the same type as one of the components that already exist in it. The added copy (or copies) must be different from the original in some way. The fact that the copy we add is different than the original component helps us exceed mundane, quantity-based ideas. The newly added copy can be different in a variety of parameters, such as color, viscosity, composition, size, or any other. On the other hand, limiting ourselves only to types of components that already exist in the system increases the probability that resulting ideas will be feasible and will thus be implemented.

Applying Multiplication

1. List the components of the product (at least 6–8).
2. Multiply one of the components by 2, 3...100, etc. without yet having an idea why this one was selected or where this action is going to lead.
3. Change the multiplied copy or copies of the component in some way that will distinguish it from the original (e.g., size, color, position, thickness, material type, and speed of action).
4. Visualize the resulting virtual product your trigger for novel thinking.
5. Identify potential opportunities, uses, benefits, markets, and advantages (the Need Filter – can we sell it?). At this stage, allow your mind to dwell exclusively on the benefits. If useless, reject the trigger and create a new one. If you have identified an opportunity, move to the next filter.
6. Apply the Viability Filter (can we make it? How?)
7. Make necessary adaptations.

Examples

- SMC is itself an example of Multiplication since the move from a combination of two materials to a compound of two resins can be seen as a Multiplication of one of the resins with a change of its properties to fiber, and thus a change in function.
- A well-known example is the multiple flashes in cameras that eliminate the red-eye phenomenon. The first flashes cause the iris to contract and thus avoid the red color of traditional flash-lighted photos.

Tips

1. The idea in Multiplication is to transcend a mere change in quantity in order to achieve a qualitative change. Thus, for example, the change from a one-bin trash-can to a two-bin one is a valid Multiplication since the new concept of garbage separation for recycling has been introduced into the category. On the other hand, the change from a two-bin to a three-bin can, while possibly being a valuable idea from a marketing point of view, is not a true application of this tool.
2. It is not always necessary to actually replicate a component; it is sometimes enough to create a "virtual" copy of it – a mirror image, for instance.
3. In a sense, this tool is the inverse of Subtraction, since instead of subtracting from a product, we add to it. However, the addition is extremely constrained, since one is only allowed to add elements of a kind already existing in the product, and one is required to change the multiplied element.

ATTRIBUTE DEPENDENCY

The basic principle underlying this tool is creating and dissolving dependencies between variables of a product, service, or strategy to create a new and innovative version. In Attribute Dependency, one works with variables rather than components (as opposed to the other tools). A variable is something that has the potential to change values; it is easy to identify as a characteristic that can change within a product or component (e.g., color, size, and material). Variables can be internal and external; internal variables are those that can be controlled by the manufacturer, while external variables interact with the product or process but are out of the manufacturer's control. In a drinking glass, for example, the thickness of the glass is internal (since it is determined by the manufacturer), while the temperature of the liquid poured into the glass is external.

Applying Attribute Dependency

1. Create a list of a product's internal and external variables (start by selecting 5–6 of each).
2. Pair off the variables, playing internal against internal and internal against external variables, by asking: does one of the variables change with the other or because of the other (i.e., is it dependent on the other?) Example: Does the length

    If there is no dependency, create one (i.e., imagine that one variable does change with or because of the other)

    If there is a dependency, change it (i.e., imagine that a different dependency or not dependency at all)
3. For each pair, identify whether a dependency already exists between the variables.
4. If a dependency does not exist, create one; if it does, try to change or dissolve it.
5. Visualize each newly created virtual product form.
6. Identify its potential benefits, markets, and advantages (market filter – can we sell it?)
7. Apply the implementation filter (can we make it?)
8. Make necessary adaptations.

Tips

1. Start with no more than 5–6 variables of each kind. Other variables will either emerge while working or can be added after these have been explored.
2. A matrix, listing internal variables on one axis and both internal and external variables on the other axis, can help organize the list and facilitate the work.
3. Pick the first pair at random, or intuitively. Then, explore any row or column that proved to be fruitful.
4. Since Attribute Dependency works with variables of the product's components, some ideas that came up through other tools will arise again.

This redundancy is acceptable because each of the tools also has unique ideas that will more readily come from its direct application.

5. Variables can be dependent in three ways: (a) in a series, (b) in a product – user-activated, and (c) in a product – automatically activated.

For example, if you look at the variables "opacity of lens" and "intensity of light in the surrounding environment," three types of product concepts can emerge:

a. In a Series: Several pairs of optical glasses with varying levels of tint in the lenses. The user can select a pair according to the intensity of external light that day.
b. In a Product – User-Activated: A pair of optical glasses to which you attach a tinted clip-on piece when it gets sunny.
c. In a Product – Automatically Activated: A pair of optical glasses that automatically changes its level of tint to adjust to changing sunlight.

CONCLUSION

The SIT method originated with five basic tools, three of which have been described in this chapter, and has been developed to apply to a wide range of uses, developing new products, solving seemingly intractable problems, coming up with novel forms of communication, rethinking and restructuring processes, generating novel concepts for strategies, and more. Being highly structured and requiring discipline, the method is often more immediately appealing to engineers than the more common "outside the box" approaches, which are mostly related to Brainstorming and its variants. These tools (Subtraction and the like), used in conjunction with a set of principles (FFF and others), are very useful for overcoming one's fixedness and thus open the door to innovation. But once this barrier has been removed, it becomes obvious that innovating within an organization requires much more than merely overcoming individuals' fixedness. In fact, achieving innovation in a repeatable and systematic way requires overcoming fixedness on two, and at times three additional levels. A group of individuals can each be innovative individually, but innovation almost always requires teamwork, and it turns out that fixedness can occur also on the level of the team, independently of the level of fixedness of each of the team's participants. That is why there is often the need for facilitation in a team effort. This facilitation can be internal or at times external. Then, when teams are working to effectively innovate, one must pay attention to fixedness at the organizational level as well. This refers to organizational dynamics and is beyond the scope of our chapter, but it would not surprise any of our readers if I mention that the set of incentives, processes, culture, and communication style of an organization (to name several of the key characteristics) can have a strong influence both on the ability of associates to innovate and on the probability that the results of this innovations will be implemented.

Back to the individual level, we hope that this chapter can serve you as a trigger to rethink your company's products and processes, and also to turn your attention to the way you and your colleagues go about thinking in your everyday practice, making the most of your existing thinking patterns while always searching for opportunities to enrich your mental toolbox.

15 Emerging New Technology of Smart Composites
Examples and Possibilities

Brian Pillay, Haibin Ning,
Mehrdad N. Ghasemi Nejhad, Probir Guha,
and Prateep Guha

CONTENTS

INTRODUCTION TO SMART COMPOSITES

Composites have become ubiquitous with high strength, low weight applications in all industrial sectors, like automotive, aerospace, mass transit, infrastructure biomedical, energy, and several other industries. It has been well established that composite materials help achieve tomorrow's technology with their flexibility in composition, manufacturing, and functionality. The next step in advancing composites is through making the material "smart," i.e., including functionality that goes beyond high strength and

DOI: 10.1201/9781003161738-15

low weight. The following are a few examples of functionalities that are either developed or under development to further advance the use of composite materials:

1. Shielding: Electromagnetic interference (EMI) can be effectively shielded by adding absorption and reflection agents to the composite laminate. Depending on the applications, composite structures are required to possess shielding functions to interference caused by different electromagnetic energy sources including radio waves, X-ray, gamma ray, etc. Some of the constituent materials used in composite structures already have the shielding functions to radiation within certain frequency ranges, which makes the composite a shielding material without any modification. For example, carbon fibers show great effectiveness in shielding the radio wave with a frequency ranging from 0.3 MHz to 1.5 GHz [1], besides their excellent structural performance. However, the modification to the composite is indispensable in order to incorporate shielding function. For instance, metal particles, such as tungsten particles, are added to polymers to induce effective shielding to X-rays [2].

2. Temperature: Thermal conductivity is becoming increasingly important and can be achieved either through conductive reinforcing agents or through manipulation of the polymer, like using graphene, etc. Thermal conductivity is an important aspect for structural composites that is used in applications like deicing and thermal imaging for defect detection. However, the use of thermal conductivity could be used as a sensor that provides signal feedback with respect to some functionality. A thermocouple is a very simple sensor that is used extensively for temperature measurement. It works simply by using two different types of metals joined at one end that when heated or cooled creates a voltage. A composite can be used in the same manner. As an example, carbon fiber itself has different resistance in the length and thickness direction due to the orientation of the graphite planes. A "thermocouple" could potentially exist at every 0/90 intersection of the carbon fiber composite. A temperature change will result in a voltage from the intersection, which will allow for either signal reception or transmittance. NGPs and CNTs can also be used in the matrix to create the same effect between the fiber and conductive paths in the matrix.

3. Self-healing refers to the function by which composite laminates automatically repair damage through different approaches, such as encapsulated monomers and catalysts, agents, solvents, and other reactive chemicals. These agents are released when the composite is damaged and will react and repair the damaged area. A typical example includes introducing some microcapsules filled with monomer and other microcapsules filled with catalyst into the epoxy matrix in a carbon fiber epoxy composite [3]. When there is any occurrence of damage, the microcapsules break and release the monomer and catalyst, both of which make contact and trigger a cross-linking reaction. The cured healing agent then repairs the damaged area by bridging cracks [3].

4. Motion/Actuation: Providing motion through differences in coefficient of thermal expansion and piezoelectric or resistive techniques to actuate sensors. It is clearly evident from the prior sections that any relative motion in the composite can result in electrical signals being generated. These electrical

Smart Composite

State awareness		Adaptiveness
Strain	**Composite**	Shielding
Stress	**Structure**	Self-healing
Temperature		NDE evaluation
Health monitoring		Energy harvesting

FIGURE 15.1 Smart composite with the functions of state awareness and/or adaptiveness. (Adapted from Reference [4].)

signals can be used as sensing devices, as in strain measurements, etc., or they could be used to actuate other sensors or devices depending on the feedback received and corrective action required. As mentioned above, the vibration in the composite structure could be measured through the interaction of the material constituents which is an indication of road conditions in an automotive. If the vibrations are above a certain threshold, high implies poor road conditions like dirt roads, the traction control on a vehicle could be enabled automatically.

Those functions of the smart composite can be categorized into state awareness and adaptiveness. State awareness functions include sensing of the material response caused by external stimuli including stress, strain, temperature, and damage (structural health monitoring). The other function, adaptiveness, of the smart composites can be change of the composite material triggered by the stimuli. Figure 15.1 shows the smart composite with the main functions of state awareness and adaptiveness.

EXAMPLES OF SUCCESSFUL INTEGRATION SMART SENSORS IN COMPOSITES & APPLICATIONS OF SMART COMPOSITES

INTEGRATION OF FIBER-OPTICS WITHIN SMART COMPOSITES

Kalamkarov et al. [5,6] introduced techniques where fiber-optic strain sensors are successfully embedded in glass- and carbon fiber-reinforced polymer (GFRP and CFRP) tendons during pultrusion. The study of the performance of the embedded Fabry–Perot fiber-optic sensors under conditions of static and dynamic loading when exposed to both low- and high-temperature extremes is presented. The experiments entailed subjecting the GFRP and CFRP tendons to sinusoidal and trapezoidal load waveforms of about 11 kN magnitude inside a temperature chamber. The temperature in the chamber was varied from −40°C to 60°C in increments of 20°C. The strain output from the embedded sensors was compared to that from externally mounted extensometers as well as to theoretical strain values. It was determined that the performance of the Fabry–Perot sensors was not affected by ambient temperatures falling within the range of −40°C to +60°C and the sensor readings conformed very well with the corresponding extensometer and theoretical readings [5,6].

In addition, Goossen et al. [7] introduced an automated process for embedding fiber-optics in woven fabric in a vacuum-assisted resin transfer molding (VARTM) process to create a smart composite structure with integrated optical conduits with less than 0.8413 optical loss. The optical fibers are embedded in the woven fabric in both warp and weft directions where connectorization is achieved without polishing and subsequently subjected to VARTM process to yield a smart composite.

Smart composites with the integration of fiber-optics can be used for temperature sensing (such as smart manufacturing of composites for structure–property relationship and residual stress and strain monitoring during the composites curing) as well as stress and strain evaluation (such as health monitoring of composites in service).

EMBEDDING PIEZOELECTRIC PATCHES WITHIN SMART COMPOSITES

Ghasemi-Nejhad et al. [8] introduced the manufacturing and testing of active composite panels (ACPs) with embedded piezoelectric patch sensors and actuators. The composite material employed was a plain weave carbon/epoxy prepreg fabric with 0.33 mm ply thickness. A cross-ply type stacking sequence was employed for the ACPs. The piezoelectric flexible patches employed were active fiber composites with 0.33 mm thickness [9]. Composite cut out layers were used to fill the space around the embedded piezo patches to minimize the problems associated with ply drops in composites. High-temperature wires were soldered to the piezo leads, insulated from the carbon substructure by high-temperature materials, and were taken out of the composite laminates employing cut out hole, molded-in hole, and embedding techniques [10]. The laminated ACPs with their embedded piezoelectric sensors and actuators were vacuum bagged and cocured inside an autoclave employing the cure cycle recommended by the composite material supplier. The Curie temperature of the embedded piezo patches should be and was well above the curing temperature of the composite materials. The capacity of the piezoelectric patches was measured before and after the cure for quality control. The manufactured ACPs were trimmed and then tested for their functionality. Vibration suppression as well as simultaneous vibration suppression and precision positioning tests, using hybrid adaptive control as well as PID Control techniques [11,12], were successfully conducted on the manufactured ACP *beams* and *plates* and their functionality was demonstrated. The advantages and disadvantages of ACPs with embedded piezoelectric sensor and actuator patches were presented in terms of manufacturing and performance [8].

Smart and Nano Composites and Structures for Structural Health Monitoring

Composite materials are increasingly being used in high-tech industrial sectors with very strict requirements, such as aeronautics, aerospace, marine, and land vehicles, as well as wind turbine power sectors. As they are highly heterogeneous and anisotropic, they have very complex damage modes. Polymer matrix composites (PMC) structures can encounter severe environmental conditions, such as temperature, humidity, or radiation, and/or various mechanical loadings such as tensile, shear, or torsion. Therefore, it is of significant importance to have a precise idea of the stress/strain rates they are experiencing as well as their damage initiation and propagation to decide when to put them out of service for maintenance or replace them if their damage is too severe [13] to avoid catastrophic failures. Real-time structural health monitoring (SHM) employing smart materials and systems in *smart composites* provides such capabilities. Such interests are in line with new trends for functionalized structures used for various applications such as energy harvesting, vibration control and positioning, and/or morphing. As the smart materials are incorporated within PMC structures, it is possible to perform their real-time SHM. *In situ* monitoring, though, has some disadvantages, such as manufacturing/embedding difficulties, or intrusiveness risks as the integrated device may degrade the mechanical properties of the smart composites [13], hence care must be taken not to degrade the properties of the host structures [5–13]. Smart materials such as fiber-optics and piezoelectric materials have been successfully embedded in PMC for various applications, without degrading the PMC host materials properties [5–13]. The external non-destructive testing techniques were implemented on the surfaces of the specimens to verify the multiphysical couplings between these external measurements and the internal capacitance curves for verifications of their SHM of the smart composites.

SPACE, AEROSPACE, AND LAND APPLICATIONS

There is a great body of research related to active vibration suppression in smart composites (or intelligent/adaptive structures). Of particular note is a body of research performed by Ghasemi-Nejhad and coworkers with *simultaneous precision positioning and vibration suppression (SPPVS)*. The technologies reviewed in this section have applications in space, aerospace, and land vehicles industries.

Ma and Ghasemi-Nejhad developed a "Frequency-Weighted Adaptive Control for Simultaneous Precision Positioning and Vibration Suppression of Smart Structures" for ACPs with embedded active fiber composite and surface mounted monolithic piezoelectric patches [32], where excellent results were achieved (see Figure 15.2).

Next, Ma and Ghasemi-Nejhad developed a "Simultaneous Precision Positioning and Vibration Suppression of Reciprocating Flexible Manipulators" and "Adaptive Control of Flexible Active Composite Manipulators Driven by Piezoelectric Patches and Active Struts with Dead Zones" using a fuzzy logic control for active composite struts (with piezoelectric stacks [14] and precision motors [15] for fine and coarse positioning, respectively) as well as active composites panels with surface mounted active fiber composite (AFC) piezoelectric patches [9] (see Figure 15.3) [16,17], where excellent results were achieved.

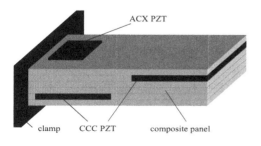

(a) schematic

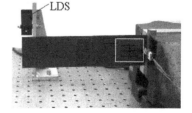

(b) photo

FIGURE 15.2   Active composite panel (ACP) [32].

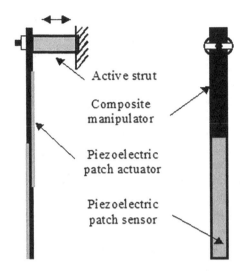

FIGURE 15.3 Flexible manipulator with active strut and piezoelectric sensor/actuator patches [17].

Ghasemi-Nejhad et al. [8] introduced a technique to integrate AFC piezoelectric patches [9] with their leads within carbon-fiber epoxy composites for SPPVS (see Figure 15.4a) [8] which in comparison with the surface mounted on the same actuators (see Figure 15.4b) [18]. They achieved excellent comparisons and verifications in terms of effectiveness of the SPPVS.

Ghasemi-Nejhad et al. also combined the ACPs with SPPVS capabilities introduced in Refs. [8] and [18] and active composite struts with SPPVS introduced in Refs. [17] and [19] developed smart composites platforms with SPPVS for both a thruster of a satellite to suppress the vibration due to thruster firing of a satellite for position keeping in orbit while at the same time precision positioning of the thruster to correct any thrust vector misalignment due to the thruster firing as well as for laser communications of two or more satellites where SPPVS is also required, which both are shown in Figure 15.8 and explained in Ref. [20]. Both platforms are composed of active composites struts with SPPVS [17] and active composites panels with SPPVS [18] to yield active composite platforms with SPPVS one for thruster vector control and the other for precision laser communications for satellites (see Figure 15.5) each with two degrees of tip-tilt freedoms [20] providing component-level nanopositioning [21,22].

SMART COMPOSITES – POSSIBILITIES IN AUTOMOTIVE

In early 2000, Probir Guha collaborated with Mike Siwajek, a contributor to this book, in an effort to incorporate surface electrical conductivity on a molded composite part. The effort was designed as a value-add to molded sheet molding compound (SMC) automotive body panels.

Automotive body panels molded from SMC all were (and still are) coated with a conductive surface primer before being sent to the automotive Original Equipment

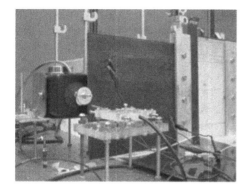
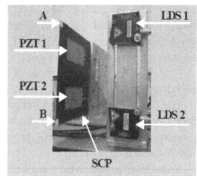

(a) Embedded Piezo Patches [24]
     (PZTs) [52]

(b) Surface mounted Piezo Patches
    LDSs are Laser Sensors

FIGURE 15.4   Active composite plates with embedded and surface-mounted piezo patches for SPPVS.

FIGURE 15.5   Active composite platforms with SPPVS for satellites: (a) Thrust vector control (on the left of the satellite in the figure) and (b) laser communications (on top of the satellite in the figure) [20].

Manufacturer (OEM) assembly plants. All topcoat applications at the OEM plant are applied using electrostatic techniques to maximize the transfer efficiency of the paint. This in turn minimizes emissions from the paint line and also improves the surface appearance of the molded part. This paint application technique is easily achieved on stamped steel since the steel provides the electrically conductive surface required for the paint application technique.

Therefore, the molded SMC body panels required a minimum amount of surface conductivity to avail of the same topcoat process and emulate similar painted surface finish. Thus, the molded SMC part had to go through the conductive primer treatment as a post-molding process.

More handling, more chances of creating defects, more cost – an all too familiar unending cycle.

The development that we worked on was to impart an adequate amount of conductivity to the molded part so that we would eliminate the need for the post-molding conductive prime process. This development has been described in a patent by Guha and Siwajek in 2007 [23].

Years later as the expertise of the textile industry is now being used to address opportunities in the composites industry, we are seeing how conductive paths and, indeed, conductive surfaces can be incorporated in molded composite parts. Devoid of the thinking and methods of the textile industry we possibly would not have seen this elegant way to incorporate unique characteristics – electrical conductivity being one – on to molded composite surfaces.

The effort and learnings from the electrically conductive polyester molding composition patent did not do much to eliminate the need for conductive priming of molded SMC parts, but we did learn a lot about handling of high viscosity moieties and this I am sure will lead to innovations in other applications. This may even lead to innovations in the use of nanoparticles in molding compounds. But the seeds of thoughts on conductivity in a molded polymer had been planted and this may have then led to the innovation of electrically conductive paths in molded components.

The manner in which conductive paths can be introduced in near-net-shape fiber preforms and thus in a molded composite product has been discussed in several patents by Guha, Han, and Pool in 2018 and 2019 [24–26]. The technology demonstrated no-touch sensors utilizing conductive paths in preforms and subsequently in molded components.

These learnings have been successfully used to demonstrate the use of conductive paths for effective sense of motion.

Now the science of no-touch motion sensing is not new and is being currently used in automotive applications. Multiple companies have used the science of "Mechatronics" to bring mechanical, electrical, and electronic skills into products for no-touch sensing and put them to use for hither-to-fore unavailable attributes for passenger safety and comfort in automobiles.

All of these methods, however, have required an additional sub-assembly that had to be married to the final assembly. This meant new complexity to the entire process – more assembly, a new item to purchase and manage, and in all likelihood more weight, probability of quality defects. A typical manufacturing process flow to incorporate smart devices in a final product using conventional technology (not smart composites) is shown in Figure 15.6.

The ability to introduce conductive paths and smart sensors into molded components gives the engineer the ability to bring in the features – sensors and actuators – described earlier in this chapter into a molded part with a decreased amount of complexity. A typical manufacturing process flow chart is shown in Figure 15.7. You can see the significant reduction in manufacturing complexity between the conventional manufacturing process in Figure 15.6 and the smart composite approach is shown in Figure 15.7.

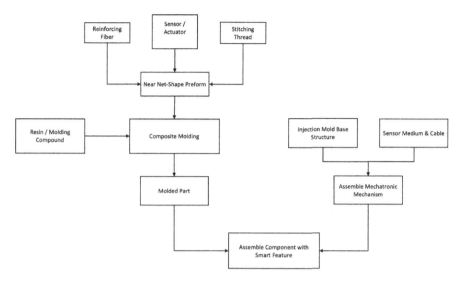

FIGURE 15.6   Component with smart feature – conventional manufacturing process steps.

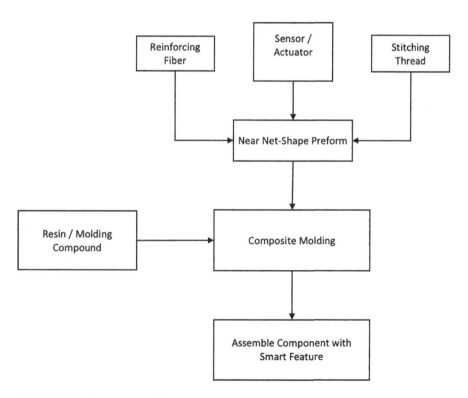

FIGURE 15.7   Component with smart feature – smart composite manufacturing process steps.

Sensors and actuators. Textile processes. Threads and yarn. Embroidered preforms. Composites.

It appears to be the right time for all of this varied knowledge base to come together in a very significant manner in the rapidly evolving field of smart composites to bring value into automotive.

Smart composites have been used successfully in many applications – civil engineering structure, satellite positions, etc. The difference here is to incorporate these features in a high-volume manufacturing process in a cost-effective manner. The innovations discussed in this section show that we are well on the way to achieve these goals.

Let us, however, also learn from our experiences with innovations in the past. The features we add to the composite process/product have to be proven out to see that key attributes of the composite – strength, stiffness, and weight – are not adversely affected. Once the features of the smart sensors and actuators have been integrated into the composite part, we must prove that manufacturing and key performance features are not degraded. Test the system and prove capability to withstand the molding process, post-molding processes like top coating when necessary, survivability in torture tests through customary heat and humidity exposure tests. A lot of this has already been completed, but we are entering a new phase of development and more will need to be done.

There is a case to make for a safety or a convenience feature in automobile closure panels – hoods, doors, decklids, and liftgates – to include "no-touch" sensor technology to keep the panels from closing when it senses that there is an interference between the fully open and the fully closed positions. Some of the sensors/actuators/ composite technologies discussed have the current capability of adding this feature in a molded component to contribute the no-touch sensor feature. This feature will need to be tested through the various application conditions – effect of paint finish type; effect of environmental, especially, wet conditions; etc.

EV Battery Temperature and Proximity to the Nearest Electric Charging Center

Data shown in Figures 15.8 and 15.9 indicate that electric vehicle (EV) battery performance and vehicle range are adversely affected at both elevated and low-temperature ambient conditions.

A temperature-sensing smart composite in the proximity of EV battery in a vehicle to actuate a temperature control device in the vehicle or the information could be transmitted to the driver so that proper corrective measures could be adopted. This is an example of a possible convenience feature. Similar features may already exist, but the example is meant to jog the imagination of the composite practitioner on what features are possible with smart composites in more cost-effective ways.

Early Detection of Onset of Failure in Critical Automotive Structural Components

The inclusion of smart electronics in molded composite components now opens the door for the early detection of failures. This feature can be expected to significantly improve the cost and convenience of keeping track of the health of critical components

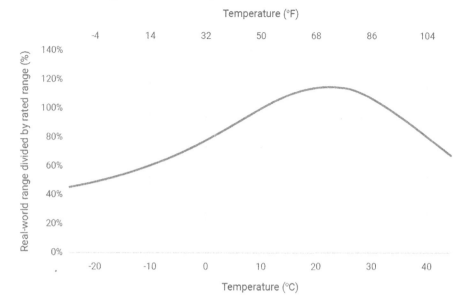

FIGURE 15.8 Effect of ambient temperature on electric vehicle range [27].

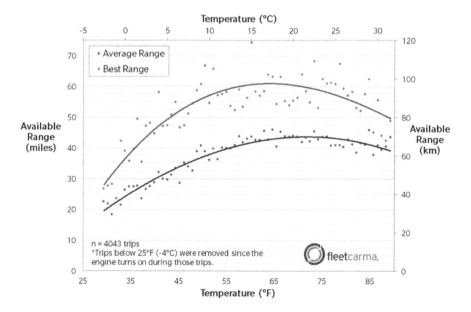

FIGURE 15.9 Effect of temperature on range of electric vehicle [28].

in a vehicle. A couple of innovations cited below demonstrates how developments in smart sensors and actuators, the use of fiber-reinforced composites in structural applications, and the capability to embroider sensors into near-net-shape preforms can all come together in a molded part to provide this feature in automotive components.

In the patent cited below, the inventors discuss the use of optical fiber sensors to measure displacement, strain, acceleration, and pressure. This has been referenced earlier in this chapter [6,7] and there have also been discussed in inventions by Feng & Dong in patents issued [29].

John Owens, GM Global Technology Operations, Inc., invention cited in an issued patent [30], demonstrates the use of fiber-reinforced composites for an anti-intrusion beam for a door assembly also known as door intrusion beam.

It is conceivable then that optical fibers (Feng patent) could be incorporated in near-net-shape preforms (Guha patent) that have been introduced to the field of automotive composites by the textile industry. This system of sensing can be incorporated in automotive structural components such as door intrusion beams (Owens patent), floor impact panels, and the like, as early detection systems for early detection of onset of failure in critical components. Here again, the capability of the composite industry to attach these sensors during the molding cycle is what provides the advantage to the user.

HEALTH MONITOR SENSORS IN AUTOMOTIVE STEERING WHEEL

The confluence of technologies – near-net-shape preforms; health monitor sensors already in play in textiles – can't be overstated. We see how they can come together to fuel new innovations through automotive applications. A patent by Vijay Varadan et al. [31] discusses textile sensors for integration into clothing to monitor a variety of outputs including neurological, cardiac, and pulmonary outputs.

The incorporation of sensors in textiles, such as clothing, bedsheets, and pillowcases, is akin to including them in near-net-shape preforms for molded composite parts. This, then, allows us to monitor health functions from a molded composite part. One immediate application for this capability would be in steering wheels of a vehicle.

The application that is possible due to parallel developments of smart sensing technologies and near-net-shape preforms are boundless. In the past, similar product attributes have been possible in molded plastic or stamped metal products, but only with an additional sub-assembly and additional post-molding/post-stamping operations. However putting together smart sensors and near-net-shape preform capability in the molded part adds a capability, a value addition that has not been widely applied to date in high-volume products. Certainly, a significant innovation to advance the composite industry.

LAUNCH COMPOSITES AND SMART SENSORS INTO THE FUTURE WITH ARTIFICIAL INTELLIGENCE

In today's world, connected automobile innovations provide automobile engineers with a wealth of information which could be harnessed for numerous benefits. By exploring the potential of connected automobiles, the author suggests vehicles which have the ability to continuously push and exchange data to and from the cloud. Thus the time series data would be made available to the Artificial Intelligence (AI) data scientist to establish relationships between these variables. These data points would include battery performance, temperature readings for various parts of the vehicle, ambient temperature, vehicle speed, RPM, miles traveled, gear and brake performances, and data from the smart composites which would include vibration readings, stress on components, roads navigated, health data from the wearable devices of the driver, and even the facial expression of the driver to ascertain their emotions, as well as any other external information such as weather information. Much of this data integration would also be contingent on addressing privacy concerns between the end-user and the scientist throughout the data collection process.

Based on data collected from trial runs, AI models could be kept ready for continuous feedback to the manufacturers and the owners of the vehicles. New metrics could be defined based on the data collected and models run. Comparison of these metrics across vehicles would also enable us to identify areas of improvement. Data and model outputs could be used to measure customer safety and comfort and provide valuable run time or batch mode inputs to customers and the manufacturers for improving vehicle performance.

While most examples cited have a heavy automotive bias, the concepts can easily translate across industry and application boundaries. In Figure 15.10, we have summarized the information flow and AI analysis possibilities

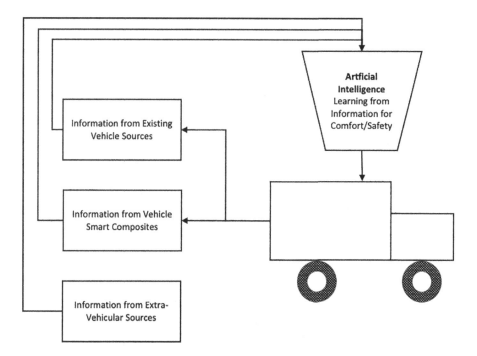

FIGURE 15.10   Smart composite–artificial intelligence interaction.

REFERENCES

1. Shui, X. and Chung, D.D.L., 2000. Submicron diameter nickel filaments and their polymer-matrix composites. *Journal of Materials Science*, *35*(7), pp. 1773–1785.
2. Kim, Y., Park, S. and Seo, Y., 2015. Enhanced X-ray shielding ability of polymer–nonleaded metal composites by multilayer structuring. *Industrial & Engineering Chemistry Research*, *54*(22), pp. 5968–5973.
3. Kessler, M.R., Sottos, N.R. and White, S.R., 2003. Self-healing structural composite materials. *Composites Part A: Applied Science and Manufacturing*, *34*(8), pp. 743–753.
4. Giurgiutiu V., 2018. 7.19 Smart materials and health monitoring of composites, Editor(s): P.W.R. Beaumont, C.H. Zweben, *Comprehensive Composite Materials II*, Elsevier, pp. 364–381.
5. Kalamkarov, A.L., and Fitzgerals, S.B., 2000, May 16. Methods for recovering leads embedded within a composite structure. US Patent No. 6,061,902.
6. Kalamkarov, A.L., Fitzgerals, S.B., MacDonald, D.O., and Georgiades, A.V., 2000, June. The mechanical performance of pultruded composite rods with embedded fiberoptic sensors. *Composites Science and Technology*, *60*(8), pp. 1161–1169.
7. Goossen, K., Wetzel, E.D., and Caruso, R.P., 2007, May 10. Automated process for embedding optical fibers in woven composites. US Patent Publication No. US 2007/0103928 A1.
8. Ghasemi Nejhad, M.N., Russ, R., and Pourjalali, S., 2005. Manufacturing and testing of active composite panels with embedded piezoelectric sensors and actuators. *Journal of Intelligent Material Systems and Structures*, *16*(4), pp. 319–334.
9. CCC, 2002. *Product Data*, Continuum Photonics, Inc., Billerica, MA.
10. Ghasemi Nejhad, M.N., and Chou, T.-W., 1990, January. Compression behavior of woven carbon fibre-reinforced epoxy composites with moulded-in and drilled holes. *Composites*, *21*(1), pp. 33–40.
11. Ma, K., Pourjalali, S., and Ghasemi Nejhad, M.N., 2002. Hybrid adaptive control of smart structures with simultaneous precision positioning and vibration suppression. *Modeling, Signal Processing and Control, Vittal S. Rao, SPIE's 9th Annual International Symposium on Smart Structures and Materials*, San Diego, CA, *4693*, 13–24.
12. Ma K., and Ghasemi-Nejhad, M.N., 2002. Development of smart composite structural systems with simultaneous precision positioning and vibration control. *Proceedings, JSME/ASME International Conference on Materials and Processing*, Honolulu, Hawaii, *1*, 396–340.
13. Tuloup, C., Harizi, W., Aboura, Z., and Meyer, Y., 2020. Integration of piezoelectric transducers (PZT and PVDF) within polymer-matrix composites for structural health monitoring applications: New success and challenges. *International Journal of Smart and Nano Materials*, *11*(4), pp. 343–369, Doi: 10.1080/19475411.2020.1830196.
14. ACX, Active Control eXperts. 2001; www.acx.com.
15. Physiks Instruments. 2001; www.physikinstrument.com.
16. Ma, K., and Ghasemi Nejhad, M.N., 2005, January. Simultaneous precision positioning and vibration suppression of reciprocating flexible manipulators. *Journal of Smart Structures and Systems*, *1*(1), pp. 13–27.
17. Ma, K., and Ghasemi Nejhad, M.N., 2008. Adaptive control of flexible active composite manipulators driven by piezoelectric patches and active struts with dead zones. *IEEE Transactions on Control Systems Technologies*, *16*(5), pp. 897–907.
18. Ma, K., and Ghasemi Nejhad, M.N., 2007. Adaptive input shaping and control for simultaneous precision positioning and vibration suppression of smart composite plates. *Journal of Smart Materials and Structures*, *16*, pp. 1870–1879.

19. Doherty, K.M., and Ghasemi Nejhad, M.N., 2005. Performance of an active composite strut for an intelligent composite modified stewart platform for thrust vector control. *Journal of Intelligent Material Systems and Structures, 16*(4), pp. 335–354.
20. Ghasemi Nejhad, M.N., 2013, March 10–14. Design of smart composite platforms for adaptive trust vector control and adaptive laser telescope for satellite applications, Active and Passive Structures and Integrated Systems VII (SSN03), H. Sodano, W.-H. Liao, and G. Park, Eds., *Proc. of SPIE's 20th International Conference on Smart Structures/NDE,* San Diego, CA, Conference 8688, Vol. SS13-SSN03-71 (Tracking No.), Paper No. SSN03-71.
21. Ma, K., and Ghasemi Nejhad, M.N., 2007. Smart composite platforms for satellite thrust vector control and vibration suppression. *Progress in Smart Materials and Structures Research,* Chapter 6, (ISBN: 1-60021-106-2, P.L. Reece, Ed.), Chapter 6, Nova Science Publishers, Inc., Hauppauge, NY.
22. Ma, K., and Ghasemi Nejhad, M.N., 2011. Smart composite systems with nanoposition-ing. Chapter 8, Handbook of Nanophysics Functional Nanomaterials, (K. Sattlar, E d.), https://www.routledge.com/Handbook-of-Nanophysics-Functional-Nanomaterials/Sattler/p/book/9781138111936.
23. Guha, P.K., and Siwajek, M.J., Patent - US20080096032- Electrically conductive polyester molding composition having a high quality surface finish.
24. Guha, P.K. Patent - US 20200101641- Vehicle component based on selective comingled fiber bundle positioning form.
25. Guha, P.K. Patent - WO 2019008444- Fiber preform of commingled fiber bundle for overmolding.
26. Guha, P.K., Han, G., and Pool, T. Patent - WO2020102363- Vehicle component based on selective commingled fiber bundle having integral electrical harness and embedded electronics.
27. https://www.geotab.com/blog/ev-range/.
28. http://ecomento.com/wp-content/uploads/2013/12/electric-car-cold-weather-range-01.jpg.
29. Feng, M.Q., and Chu, D. Patent -5969342- Multiplexable optical fiber displacement, strain acceleration and pressure sensors and method of operating the same.
30. Owens, J.N. Patent -20100266806- Anti-intrusion beam for vehicle door assembly.
31. Varadan, V., Rai, P., Kumar, S., Mathur, G., and Agarwal, M.P. Patent -20130211208- Smart materials, dry textile sensors, and electronics integration in clothing, bed sheets, and pillowcases for neurological, cardiac and/or pulmonary monitoring.
32. Ma, K., and Ghasemi Nejhad, M. N., 2004. Frequency-weighted adaptive control for simultaneous precision positioning and vibration suppression of smart structures. *Journal of Smart Materials and Structures, 13*(5), pp. 1143–1154.

16 Nanocomposites
Influence of Nanotechnology on Composites

*Vamshi Gudapati, Gajendra Pandey, and
Mehrdad N. Ghasemi Nejhad*

CONTENTS

INTRODUCTION TO NANOTECHNOLOGY

Nanotechnology is simply science and engineering carried out on the nanometer scale, that is, 10^{-9}m. Nanotechnology had its origins in processes used to create gold and silver coatings for creating stained glass windows hundreds of years ago. Similarly, chemists have been making polymers made up of nanometer-sized molecules for decades. However, the ideas and concepts behind nanotechnology started with a talk "There's Plenty of Room at the Bottom" by physicist Richard Feynman long before nanotechnology was used [1a]. Nanotechnology can be broadly defined as the creation, processing, characterization, and utilization of materials, devices, and systems with at least one dimension in the order of 0.1–100nm [2a]. To exploit the benefits of nanotechnology, one needs the ability to maneuver at the scale of atoms and molecules and to thoroughly characterize material properties at this scale. The ability to manipulate matters at the atomic-, molecular-, and macromolecular

DOI: 10.1201/9781003161738-16

level has led to the discovery of nanomaterials that have significantly different and superior properties compared with the same material at the bulk scale.

When a material is in the visible range with the human naked eye, the material properties do not change significantly with size. However, properties of nanomaterials change significantly from those at larger scales. Nanomaterials have relatively large surface-to-volume ratios and are governed by quantum effects that rule behaviors and properties of particles at nanoscale. Due to change in scale and associated governing laws, nano-size materials exhibit novel and significantly different and superior physical, chemical, optical, electrical, magnetic, and biological properties. For example, particles at nanoscale exhibit melting point, fluorescence, electrical and thermal conductivity, magnetic permeability, and chemical reactivity changes as a function of the size of particles. That is, a material property of interest can be fine-tuned by changing the size of particles. Figure 16.1 shows "The Scale of Things—Nanometers and More" (courtesy of US DOE).

True potentials of nanotechnology can be exploited using multidisciplinary efforts at nanoscale in chemistry, biology, physics, materials science, and engineering. In addition to cross-disciplinary efforts, nanotechnology in modern era is being enabled by (1) improved control and manipulation of nanoscale materials, (2) improved metrology tools for characterization of materials at nanoscale, and (3) improved understanding

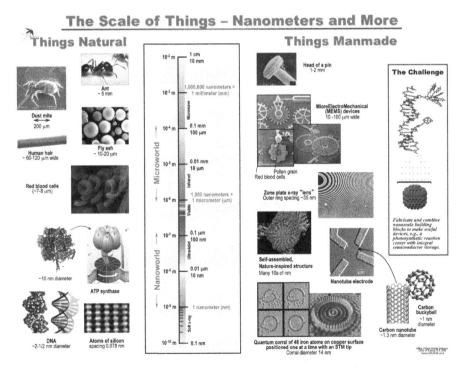

FIGURE 16.1 The size of nanoscale objects and phenomena compared with the size of small everyday objects. (Courtesy of the Office of Basic Energy Sciences, Office of Science, U.S. Department of Energy, DOE.)

of nanostructure and material property relationships at nanoscale. Potential applications of nanotechnology are boundless. Using nanocrystalline materials mixed in with traditional materials to create nanocomposites will allow new materials with novel properties to be realized, including completely stain proof cloths, armors only a few millimeters thick, televisions of any size less than a millimeter thick, completely scratch-proof optics, durable fuel cells, high-efficiency solar cells, high energy/power density batteries and supercapacitors, light-weight nanocomposites, and many others. Nanocomposites will create new products, which have not yet been considered.

NANOMATERIALS AND NANOSTRUCTURES

Nanomaterials and nanostructures may be in different shapes and sizes and may consist of various types of elements with different properties. Nanomaterials can be classified into carbon-based nanomaterials, metal and metal oxide-based nanomaterials, organic-based nanomaterials, and composite-based nanomaterials called nanocomposites. Common nanomaterials used in nanocomposites are nanoparticles, nanoclays, and carbon-based nanomaterials. Carbon-based nanomaterials contain carbon in morphologies such as spheres, ellipsoids, hollow tubes, rods, or sheets also known as fullerenes, carbon nanotubes (CNTs), carbon nanofibers, and graphene (see Figure 16.2), also called "allotropes of carbon."

Among the many different types of nanomaterials, CNTs and graphene nanosheets (GNSs) have been the most interesting nanostructures ever introduced due to their superior mechanical, thermal, electrical, optical, and chemical properties. CNTs and GNSs have tensile modulus and strength values ranging from 270 GPa to 1 TPa and 11–200 GPa, respectively [20]. Macroscale graphite flakes with a thickness of 0.4–60 mm may expand up to 2–20,000 mm in length. These sheets/layers can be separated down to 1 nm thickness, forming a high aspect ratio (200–1,500) and high modulus (~1 TPa) GNSs. Furthermore, when dispersed in the matrix, the nanosheet exposes an enormous interface surface area (2,630 m^2/g) and plays a key role in the improvement of both physical and mechanical properties of the resultant nanocomposite.

NANOCOMPOSITES

Nanocomposites are of significant importance in the rapidly developing field of nanotechnology. Researchers have investigated composites containing polymer nanoparticles to improve their physical, mechanical, and chemical properties. Nanoparticles

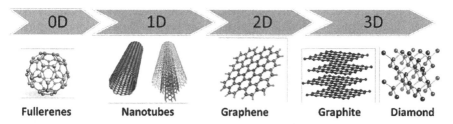

FIGURE 16.2 Carbon-based materials with different morphologies (https://inscx.com/wp-content/uploads/2018/03/Crystal-structure-of-carbon-nanomaterials-1.png.)

embedded in a polymer matrix have attracted increasing interest because of the unique properties displayed by nanoparticles and their inclusion in polymers. Due to the nanometer size of these particles, their physicochemical characteristics differ significantly from those of micron size and bulk materials. When two or more phases are mixed together to make a composite, one can often obtain a combination of properties, in the resulting composite, that are not available in either one constituent.

The composites/nanocomposites of the future will offer many advances over composites of today. Recent developments in the production and characterization of various nanoparticles have created numerous new opportunities to develop nanocomposites for a variety of different applications. The field of nanocomposites involves the study and development of multiphase materials where at least one of the constituent phases has one dimension less than 100 nm. This is the range where the phenomena associated with the atomic and molecular interaction strongly influence the macroscopic properties of materials. Since the building blocks of nanocomposites are at nanoscale and nanomaterials have enormous surface areas as compared to their volumes, there are large and numerous interfaces and interphases between the two intermix material phases. The special properties of the nanocomposite arise from the interaction of its phases at the interface and/or interphase regions. By contrast, in a conventional composite based on micrometer-sized fillers such as carbon fibers, the interfaces between the filler and matrix constitutes have a much smaller surface-to-volume ratio and hence influence the properties of the host structure to a much smaller extent. The optimum amount of nanoparticles in the nanocomposites depends on the filler size, shape, homogeneity of particle distribution, and the interfacial bonding properties between the fillers and the matrix. The promise of nanocomposites lies in their multifunctionality, i.e., the possibility of realizing a unique combination of properties unachievable with traditional materials. The challenges in reaching this promise are tremendous. They include control over the distribution in size and dispersion of the nano-size constituents, and tailoring and understanding the role of interfaces between physically and chemically dissimilar phases on bulk properties.

Scientists and engineers working with fiber-reinforced composites have practiced a system of "bottom-up" approach in processing and manufacturing, at micron level, for decades. When designing a composite, the material properties are tailored for the desired performance across various length scales. From the selection and processing of matrix and fiber materials and architecture to the layup of laminae in laminated composites, and net-shape forming of the macroscopic composite parts, and hence the integrated approach used in composites processing is a remarkable example in the successful use of the "bottom-up" approach (even prior to the development of nanocomposites—albeit at the micron level). Conventional composites are manufactured using polymers combined with fibers. The integration/combination of various nanomaterials with various resins gives nano-resin nanocomposites with multifunctionality [1–6]. When these nano-resin nanocomposites are combined with fibers, hierarchical nanocomposites are produced with multifunctionality. On the other hand, when CNTs are grown on the surface of fibers, nanoforest (NF) fibers are produced [7–9]. When resins are combined with NF fibers, multifunctional hierarchical nanocomposites are produced [10,11]. Figure 16.3 gives the classification of nanocomposites [20].

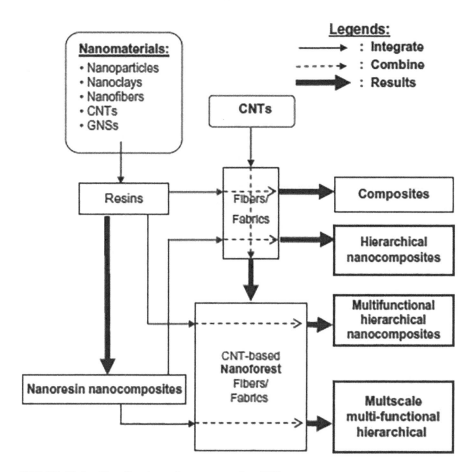

FIGURE 16.3   Classifications of nanocomposites [20].

MANUFACTURING AND PROCESSING OF NANOCOMPOSITES

The composites and nanocomposites of the future will offer many advances over the composites of today. Recent developments in production and characterization of various nanoparticles and nanomaterials have created numerous new opportunities to develop nanocomposites for a variety of different applications. The potential to develop nanoparticle-reinforced nanocomposites looks promising for a wide range of applications including high mechanical damping, strength, fracture toughness, and electrically and thermally conductive polymer nanocomposites [1–6]. However, the application of nanomaterials as structural reinforcements depends on the ability of nanomaterials to successfully transfer load from the matrix through proper interfaces and interphases. To effectively transfer load from nanomaterial reinforcement(s) to the resin system, nanomaterials need to be beneficially manufactured, functionalized, and processed within the base resin system. Process steps of utmost importance in preparation of nanocomposites are exfoliation, functionalization, dispersion, and optimum loading in the base resin system. Dispersion of nanomaterials into a polymer matrix has been one of the main challenges to date due to the aggregation of nanomaterials as a result of the Van der Waals forces and interactions between the nanomaterials. Uniform dispersion of nanomaterial throughout the resin leads to consistent load transfer from resin to nanomaterials and vice versa. SWCNTs and thinner GNSs with a few layers agglomerate more easily than their multiwalled or multilayered counterparts due to their size differences (i.e., relatively greater surface areas exist for thinner nanomaterials). On the other hand, the SWCNTs and thin GNSs with a few layers, when dispersed well, have been found to demonstrate higher mechanical, electrical, and thermal properties. Researchers have used many different techniques in an attempt to disperse nanotubes (such as CNTs) and nanosheets (such as GNSs) in polymer matrices, including solution chemistry to functionalize the nanomaterials surfaces, the use of polymers to coat the nanomaterials surfaces, ultrasonic dispersion in a solution, and the use of surfactants. Functionalization allows nanomaterials to bond better to the resin and overcome the Van der Waals interactions between nanomaterials, yielding better interfaces and interphases. Improper dispersion leads to agglomeration of nanoparticles/nanomaterials and the agglomerated inclusions act as defects with high stress concentrations often degrading properties instead of acting as uniform reinforcements enhancing the properties.

UNMET NEEDS IN COMPOSITES

Traditionally engineered commonly used composites (i.e., primarily thermosetting-based polymer composites) suffer from two main drawbacks: (1) *Brittleness* of the resin and (2) *Lack of Through-the-Thickness Reinforcement* due to the two-dimensional (2D) nature of the basic building block of the composites, i.e., 2D plies of laminae, short, or long fibers oriented in the 2D plane. Therefore, the resulting traditional composites will have fibers in the in-plane directions (and not through-the-thickness direction) in addition to having a brittle matrix in between the plies. Conventional matrix toughening mechanisms and through-the-thickness reinforcements (such as stitching) increase the toughness and some through-the-thickness properties but adversely affect other properties (primarily the in-plane properties). When traditional 2D composites are subjected to loadings, they develop interlaminar normal and shear stresses, and when those stresses in the matrix region joining the two adjacent plies (i.e., within the weakest link of composites and where there is a matrix-rich area), reaches the normal and shear strength of the matrix (not the composite), cracks initiate in the matrix region and upon further loading such cracks can grow and when the cracks lengths reach their critical lengths a failure will happen, called "delamination." Therefore, to prevent the cracks to initiate in the weakest link (i.e., in the matrix), there are two primary remedies: (1) To convert the matrix from a brittle matrix into a tough matrix so that the cracks cannot easily initiate, and even if they do, a tough matrix will create a crack-tip blunting mechanism which will prevent or delay the crack growth/propagations. (2) To develop reinforcements within the weakest link region (i.e., the matrix-rich region in between the plies) such that even if cracks initiate, they cannot grow freely without any resistance. This section will discuss potential solutions for the development of tough and three-dimensional (3D) nanocomposites employing various nanomaterials with delamination resistance and multifunctionality to resolve both of the fundamental issues of (1) and (2) within commonly used composites as explained above.

To remedy Problem (1), i.e., the *Brittleness* of the matrices, their toughening is required. Ghasemi-Nejhad and coworkers have developed a nano-resin Technology [1–4] where a brittle resin can be converted into a tough resin by employing only about 0.02%–0.04% (by weight) of a functionalized GNSs to improve the fracture toughness of the resin by 200% [4]. The very low percentage of the nanomaterials inclusions and a substantial amount of fracture toughness improvement make such approaches in matrix toughening economical for large-scale industrial use. Toughening of brittle thermosets using nanomaterial technologies are detailed in the "Nano-resin" section in the following.

To remedy Problem (2), i.e., the *2D nature* of the laminated composites, their interply regions (i.e., the weakest link in between the layers) should be reinforced with CNTs NF (i.e., NF Technology [7–9]) to (1) fill the weakest link region of the interply gap (filled by matrix only) by CNTs NF reinforcement and (2) provide reinforcement in the through-the-thickness direction, i.e., the third direction of the laminated composite to render the resulting three-dimensional nanocomposite. This can be achieved by growing CNT nanostructures on the fibers/fabrics directly (also called NF I Technology [7–9]) to produce three-dimensional multifunctional nanocomposites. These enhancements in nanocomposites are further elaborated in the "Nanoforest" section in the following.

NANO-RESIN TECHNOLOGY

Nanomaterials are effective in improving strength of thermoplastics and toughness of brittle thermosets [1–4]. Over the past decade, the mechanical and thermal properties of thermoplastic and thermoset nanocomposites have been investigated extensively. Thermoplastic composites are widely used in automotive interior and nonstructural components. Thermosets are widely used as adhesives, coatings, and binders for structural composites. Thermoset resins such as epoxies, polyesters, and vinyl esters exhibit a high degree of cross-linking enabling them to exhibit high stiffness and strength. The highly cross-linked structure in thermosets also makes them brittle and susceptible to cracks, limiting their application in automotive parts. It is of utmost importance to improve the fracture toughness of thermoset resins. Traditionally applied tougheners such as thermoplastic inclusions and rubbers can increase toughness by forming micrometer-sized secondary phases and cause a noticeable reduction in stiffness, strength, and glass transition temperatures. Inorganic nanomaterials such as metal oxides and nanoclays somewhat improve toughness, stiffness, strength, and glass transition temperature of thermosets. However, the toughness improvements are less than polymeric inclusions.

Graphene has recently emerged as a promising material for enhancing toughness of polymer nanocomposites. It is a technical challenge to achieve full dispersion and exfoliation of nanomaterials, such as nanoclays and GNSs, due to large lateral dimensions of the layers, high intrinsic viscosity of polymer resins (especially, when a large percentage of nanomaterials are used), and a strong tendency of nanomaterials to agglomerate. With a goal of industrializing nanocomposites, we sought to investigate toughening effects of graphene nanosheets at relatively low loadings via solvent-free incorporation into thermoset epoxies. Graphenes, graphene oxide, and surface-modified graphene oxide were employed to demonstrate toughening effects of graphene nanocomposites at loadings less than 0.1 wt%. Interestingly, the maximum toughening effect was observed at 0.02 or 0.04 wt% of graphene loading for all composite samples [4]. This study demonstrated 1.5-fold improvements in fracture toughness and 2.4-fold improvements in fracture energy at 0.04 wt% of graphene loading (see Figure 16.4). Due to significant improvements in toughness at miniscule loadings of functionalized graphene, this approach is an economically and commercially viable option.

A "Crack Coalescence Failure Mechanism" was introduced in Ref. [4] to explain the behavior of the resulting nanocomposites at various functionalized GNSs loading of Figure 16.4.

It was also shown that a small amount of functionalized GNSs (about 0.06% by weight) within the filaments of a 3D polymer printer could improve the toughness of the resulting 3D printed parts [5,21]. Other toughening materials such as SiC nanoparticles and nanodiamonds were used to demonstrate an increase in fracture toughness of thermoset resin systems [6]. The very low percentage of the nanomaterials inclusions and a substantial amount of fracture toughness improvement introduced in "Nano-resin Technology" [1–6] make these approaches in matrix toughening economical for large-scale industrial mass production.

The Nano-resin Technology [1–6] resolves Problem (1) of the traditional composites (as explained above).

NANOFOREST TECHNOLOGY

Through nanotechnology, it is envisioned that nanostructured materials will be developed using a bottom-up approach, where materials and products are made from the bottom-up, i.e., by building them from atoms and molecules to nanomaterials, and micro-materials to produce macro-materials and structures [7–9]. A "hierarchical composite" is a composite with constituents that increase in size hierarchically, i.e., a nanostructure material, such as nanoparticle, nanoclay, nanotube, and or nanosheet,

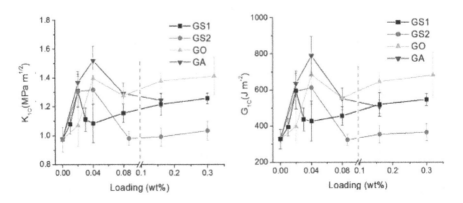

FIGURE 16.4 Fracture toughness and fracture energy of graphene epoxy nanocomposites [4].

is manufactured first. Next, this nanomaterial is mixed with a polymer to produce a nanocomposite using nano-resin technology described above. In this section, we discuss employing a bottom-up approach to *in situ* manufacture nanoscale materials (such as MWCNTs) onto microscale materials (such as carbon, glass, Kevlar, Spectra, or SiC fibers) and finally combining these with resins to give multiscale hierarchical multifunctional nanocomposites.

For many decades, advanced composites have been used as viable primary load-bearing structures. Although the in-plane loading and stresses have been handled by various configurations of fiber architectures, such as 1D (i.e., unidirectional tapes) and 2D (i.e., woven fabrics), the inter- and intralaminar stresses have remained major issues, resulting in relatively weak interlaminar fracture toughnesses. To solve this problem, Ghasemi-Nejhad and coworkers [7–9] have demonstrated controlled growth of CNTs on silicon carbide fibers using chemical vapor deposition technique to demonstrate nano-brush nanocomponent for multifunctional applications as shown below in Figure 16.5 [22].

This work formed the basis for NF I Technology [7–9] where CNT-based NF were grown directly on the surface of the fibers/fabrics (of various fiber materials such as carbon, glass, Kevlar, ceramic, etc.) to provide reinforcement in between the composite plies (i.e., in the weakest link matrix-rich regions) and act as the third direction (i.e., through-the-thickness) reinforcements [7–9].

Figures 16.6 and 16.7 detail the manufacturing steps of NF I Technology [7,8].

The 3D nanocomposites fabricated with nanotube NFs shown in Figures 16.6 and 16.7 demonstrated remarkable improvements in the interlaminar fracture toughness, hardness, delamination resistance, in-plane mechanical properties, damping, thermoelastic behavior, and thermal and electrical conductivities making these structures truly multifunctional as shown in Table 16.1. The vertical arrays of NF nanotubes in the thickness direction of the resulting nanocomposites improve the out-of-plane through-the-thickness mechanical properties without compromising the in-plane properties [7,8].

Table 16.1 shows multifunctional properties of the 3D hierarchical multifunctional NF nanocomposite as compared with their 2D woven composite counterpart.

When the CNT-based NFs are grown on the surface of a substrate and then separated and placed in between the plies of composites to produce three-dimensional hierarchical multifunctional nanocomposites, the NF technology is called NF II

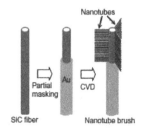

FIGURE 16.5 Illustration of masking on SiC fibers to grow CNTs on fiber top; As-grown nanotubes on top of SiC fibers creating multiple multifunctional nano-brushes.

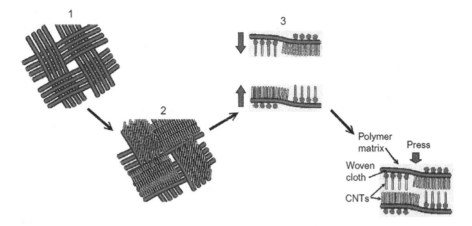

FIGURE 16.6 Schematic diagram of the steps involved in the hierarchical nanomanufacturing of multifunctional 3D nanocomposites. (1) Aligned nanotubes grown on the fiber cloth. (2) Stacking of matrix-infiltrated MWCNT-grown fiber cloth. (3) 3D MHN laminated plate fabrication by layup technique.

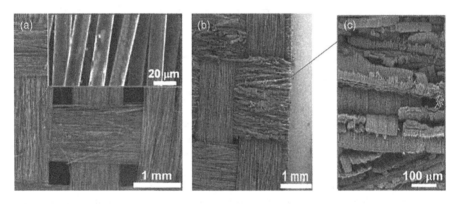

FIGURE 16.7 Growth of MWCNTs on the SiC woven cloth: (a) plain-weave SiC fabric cloth. (Inset: individual bare fibers of the woven cloth). (b) The cloth with MWCNTs grown perpendicularly on its surface. (c) Close-up view of the MWCNTs grown on the SiC woven cloth.

[12–14]. Due to superior properties of the CNTs, the NF technology developed gives multifunctionality to the resulting nanocomposites. The developed nanocomposites exhibit an increase in the structural (i.e., strength, stiffness, strain-to-failure, overall toughness, fracture toughness, and structural damping) and nonstructural (i.e., thermal and electrical conductivities as well as lowering coefficient of thermal expansions, CTE) properties. It should be noted that while NF I Technology is suitable for the wet processing of composites (since the NF I has to be grown on the surface of the fibers first), NF II technology is suitable for both wet and prepreg processing of composites/nanocomposites.

The NF Technology [7–14] resolves Problem (2) of the traditional composites (as explained above).

TABLE 16.1
Properties Enhancements of the Resulting 3D Hierarchical Multifunctional NF Nanocomposite as Compared with Their 2D Woven Composite Counterpart

Material Property Type		2D Composite	3D Nanocomposite
Fracture toughness test results	$G_{IC}(KJ/m^2)(DCB)$	0.95	4.26 (348% enhancement)
Flexure test results	$G_{IIC}(J/m^2)(ENF)$	91	140 (54% enhancement)
	Flexural modulus (GPa)	23.1 ± 0.3	24.3 ± 0.2 (5% enhancement)
	Flexural strength (MPa)	62.1 ± 2.1	150.1 ± 1.4 (140% enhancement)
	Flexural toughness (Nmm)	5.8	30.4 (424% enhancement)
Structural dynamic properties	ζ (damping ratio)	0.0095	0.0731 (669% enhancement)
	f_n (Hz) (natural frequency)	753.9	601.4
	$f_n\zeta$ (damping characteristic)	7.162	43.963 (514% enhancement)
Average CTE (a) over 0–150°C (ppm/°C)		123.9 ± 0.4	47.3 ± 0.3 (62% enhancement)
Through thickness thermal conductivity at 125°C (W/mK)		0.33	0.50 (51% enhancement)
Through-thickness electrical conductivity (S/cm)		0.075 × 10E−6 (insulating)	0.408 (conductive)

In addition, it was shown that a combination of Nano-resin and NF Technologies will improve the properties of the resulting 3D hierarchical multifunctional nano-composites more than the individual improvements due to either the Nano-resin Technology alone or the NF Technology alone when the percentage of the nanomaterials within the nano-resin and the length and population of CNTs within the NF are optimized [11,15–19].

CONCLUSIONS

Conventional composite materials in the past have traditionally suffered from two main issues: first, *brittleness* of the matrix and second, *lack of reinforcement in through-the-thickness direction in-between the plies* making them susceptible to *delaminations*. If these two primary issues can be resolved to a large extent, the applications of composite materials in *automotive, space & aerospace, and wind energy* industries will substantially increase. It is shown in this chapter that it is possible to resolve composites *brittleness* issues to a large extend using *nano-resin* technology explained here to improve the *toughness* of the resin and hence composite systems. If the matrices in composites can be toughened, the fracture toughness and damage tolerance of composites will increase, resolving the brittleness issue, and hence cracks initiations and growths in the matrices can be prevented or delayed leading to higher performance composites/nanocomposites. It is also shown in this chapter that it is possible to resolve the issues associated with the *lack of reinforcement in between the layers and through-the-thickness* in composites to a large extend using *NF* technology explained here. In a conventional commonly used composite material, in between the layers are filled by an often brittle matrix without a reinforcement, and hence it constitutes the weakest link in composites and the primary source of *delaminations*—an often mode of failure in composites. If a NF "reinforcement" is placed in-between the composite layers, then the weakest link will also be reinforced and hence the crack initiation and growth in between the layers will be eliminated or delayed to eliminate or improve the delamination susceptibility of the composites. It is conceivable that with the use of *Nano-resin* and/or *NF* Technologies, either individually or collectively, the *fatigue performance* of the resulting nanocomposites will also substantially improve due to the suppression of crack initiation and/or growth that would otherwise lead to delamination and parts failure when the critical crack length is reached.

REFERENCES

1a. Feynman, R.P. "There's plenty of room at the bottom." *Engineering and Science*, Vol. 23:5, February 1960.
2a. Ajayan, P.M., Schadler, L.S., and Braun, P.V. *Nanocomposite Science and Technology*, Wiley-VCH, Weinheim, 2003.

NANORESIN TECHNOLOGY:

1. Ghasemi Nejhad, M.N., Veedu, V.P., Yuen, A., and Askari, D., "Polymer matrix composites with nanoscale reinforcements," US Patent 2007/0142548 A1, US Patent Issued, February 9, 2010 (International Publication Number: WO/2008/060294); US Patent Number 7,658,870 B2 (Composition of Matter).
2. Ghasemi Nejhad, M.N., Veedu, V.P., Yuen, A., and Askari, D., "Polymer matrix composites with nanoscale reinforcements," US Patent 2007/0142548 A1, US Patent Issued, January 25, 2011 (International Publication Number: WO/2008/060294); US Patent Number 7,875,212 B2 (Method of Manufacturing).
3. Russ, R., Ghasemi Nejhad, M.N., Tiwari, A., Chaturvedi, A., Hummer, D. A., and Gudapati, V. "Nanomaterial-reinforced resins and related materials," (Composition of Matters & Method of Manufacturing), World-wide/International Patent Applications, PCT/US2011/030080 (Filing Date: March 26, 2011), International Publication Number: WO/2011/120008 (Publication Date: Sept. 29, 2011), US Patent "9,120,908B2" Granted on Sept. 1, 2015; Japanese Patent "JP5841125B2" Granted on Jan. 13, 2016, Chinese Patent "CN103108905A1" Granted on Jan. 13, 2016, Also filed in Canada (2794482A1), EP Countries (2553007A1), etc.
4. Park, Y.T., Qian, Y., Chan, C., Suh, T., Ghasemi Nejhad, M., Macosko, C., and Stein, A., "Epoxy toughening with low graphene loading." *Advanced Functional Materials*, Vol. 25, No. 4, Jan. 28, 2015, pp. 575–585.
5. Yamamoto, B.E., Trimble, A.Z., Minei, B.M., and Ghasemi Nejhad, M.N., "Development of multifunctional nanocomposites with 3d printing additive manufacturing and low graphene loading." *Journal of Thermoplastic Composite Materials*, Apr. 3, 2018, pp. 1–26.
6. Gudapati, V., and Ghasemi Nejhad, M.N., "Use of nanoparticles for the development of high-performance nanoresins," SAMPE Europe, *31st International Technical Conference & Forum on Improvement of Materials and Application Characteristics*, April 12–14, 2010, Paris, France.

NANOFOREST TECHNOLOGY:

NANOFOREST I (NF I) IN POLYMER COMPOSITES:

7. Ghasemi Nejhad, M.N., Veedu, V.P., Cao, A., Ajayan. P.M., and Askari, D., "Three-dimensionally reinforced multifunctional nanocomposites," World-wide/ International Patent, PCT/US2006/047483, International Publication Number: WO/2008/054409, European & Hong Kong Patent No. 1 966 286 B1 (Granted on September 30, 2009); Mexican Patent No. 279045 (Granted on September 14, 2010); Malaysian Patent No. 143396-A (Granted on May 13, 2011); Russian Patent No. 2423394 (Granted on July 10, 2011); U.S. Patent No. 8,148,276 B2 (Granted on April 3, 2012); Chinese Patent No. ZL200680051863 (Granted on April 24, 2012); Japanese Patent No. 5037518 (Granted on July 13, 2012); South Korean Patent No. 10–2008–7015810 (Granted on Oct. 2, 2013), Canadian Patent No. 2,632,202; (Granted on May 1, 2014); Other Granted Patents: Australian (2006350255/2006), Indian (5493/DELNP/2008).

8. Veedu, V.P., Cao, A., Li, X., Ma, K., Soldano, C., Ajayan, P.M., and Ghasemi-Nejhad, M.N., "Multifunctional composites using reinforced laminae with carbon-nanotube forests." *Nature Materials*, Vol. 5, Jun. 2006, pp. 457–462.

9. Askari, D., and Ghasemi Nejhad, M.N., "Effects of vertically aligned carbon nanotubes on shear performance of laminated nanocomposites bonded joints." *Journal of Science and Technology of Advanced Materials*, Vol. 13, No. 4, Jul. 15, 2012, p. 045002, doi:10.1088/1468-6996/13/4/045002.

22. Cao, A., Veedu, V.P., Li, X., Yao, Z., Ghasemi-Nejhad, M.N., and Ajayan, P.M., "Multifunctional brushes made from carbon nanotubes." *Nature Materials*, Vol. 4, July 2005, pp. 540–545.

NANOFOREST I (NF I) IN CERAMIC COMPOSITES:

10. Goodman, W.A., Ghasemi Nejhad, M.N., Wright, S., and Welson, D. "T300HoneySiC: A new near-zero CTE molded C/SiC material," *SPIE Optical Engineering and Applications Conferences, Material Technologies and Applications to Optics, Structures, Components, and Sub-Systems II*, Paper: SPIE 9574-13, Aug. 9–13, 2015, San Diego, CA, Conference 9574, Paper No. SPIE 9574-13.

11. Khosroshahi, K., and Ghasemi Nejhad, M.N., "Processing and characterization of nanoparticles and carbon nanotube reinforced continuous fiber ceramic nanocomposites by preceramic polymer pyrolysis," *SPIE Conference on Optomechanics and Optical Manufacturing; Material Technologies and Applications to Optics, Structures, Components, and Sub-Systems IV* (Conference OP304); Chairs: Matthias Krödel; William A. Goodman; Aug. 11–15, 2019, San Diego, CA, Ref. No. OP304-18, Paper 11101-4 (Invited Paper), pp. 1–13.

NANOFOREST II (NF II):

12. Cao, A., Dickrell, P.L., Sawyer, W.G., Ghasemi-Nejhad, M.N., and Ajayan, P.M., "Super-compressible foam-like films of carbon nanotubes," *Science*, Vol. 310, Nov. 2005, pp. 1307–1310.

13. Ghasemi Nejhad, M.N., and Gupapati, V., "Nanotape and nanocarpet materials," World-wide/International Patent, PCT/US2011/020360; Filing Date: Jan. 6, 2011, WO Number: WO2011106109A3; WO Publication Date: Jan. 9, 2011; US Patent Publication Number: US 2013/0216811 A1 (Aug. 22, 2013), EP Patent Publication No. 2521694/11729194 (Nov. 14, 2012).

14. Gudapati, V., and Ghasemi Nejhad, M.N., "Multifunctional adhesive joints for composites using aligned carbon nanotube nanofoam films," *Aerospace Division, Adaptive Structures and Material Systems, ASME Smart Materials, Adaptive Structures and Intelligent Systems Conference, Multifunctional Materials Symposium*, Oct. 28–30, 2008, Ellicott City, MD, Paper No. SMASIS08-586.

HIERARCHICAL MULTIFUNCTIONAL
NANOCOMPOSITES – CONFERENCE PAPERS:

15. Ghasemi Nejhad, M.N., Cao, A., Veedu, V.P., Askari, D., and Gudapati, V., "Nanocomposites and hierarchical nanocomposites development at Hawaii nanotechnology laboratory," *ASME First International Conference on Multifunctional Nanocomposites 2006*, ASME Nanotechnology Institute, Sept. 20–22, 2006, Honolulu, HI, Paper No. MN2006-17053.
16. Ghasemi Nejhad, M.N., and Cao, A. "Three dimensional multifunctional hierarchical nanocomposites-multifunctional materials," Behavior and Mechanics of Mutifunctional Composites, Marcelo J. Dapino & Zoubeida Ounaies, *Proc. of SPIE's 14th International Symposium on Smart Structures and Materials*, March 18–22, 2007, San Diego, CA, Vol. SS07-SSN04-78.
17. Ghasemi Nejhad, M.N., "Multifunctional hierarchical nanocomposites – A review," ASME International Mechanical Engineering Congress and Exposition; Advanced Manufacturing; Track 10: Mechanics of Solids, Structures, and Fluids; Topic14: Hierarchical Nanocomposites, Nov. 15–21, 2013, San Diego, CA, Paper No. IMECE2013-65599.
18. Ghasemi Nejhad, M.N., "Hierarchical multifunctional nanocomposites," *Behavior and Mechanics of Multifunctional Materials and Composites VIII (SSN04)*, (N.C. Goulbourne and H.E. Naguib, Eds.), *Proc. of SPIE's 21st International Conference on Smart Structures/NDE*, March 9–13, 2014, San Diego, CA, Conference 9058, Vol. SS14-SSN04-27 (Tracking No.), Paper No. SSN04-9058-7.
19. Ghasemi Nejhad, M.N., "Applications of nano and smart materials in renewable energy production and storage devices," *Smart Materials and Nondestructive Evaluation for Energy Systems (SSN11)*, (N.G. Meyendorf and T.E. Matikas, Eds.), *Proc. of SPIE's 22nd International Conference on Smart Structures/NDE*, March 8–12, 2015, San Diego, CA, Conference 9439, Vol. SS15-SSN11-21 (Tracking No.), Paper No. SSN11-21_9439-17.

HIERARCHICAL MULTIFUNCTIONAL
NANOCOMPOSITES – BOOK CHAPTERS:

20. Ghasemi Nejhad, M.N., "Multifunctional hierarchical nanocomposite laminates for automotive / aerospace applications," (Chapter 15, pp. 491–526) *Multi-Functionality of Polymer Composites: Challenges and New Solutions – Applications in Transportation*, (K. Friedrich and U. Breuer, Eds.), Elsevier Publications, Waltham, MA, USA, 2015.
21. Trimble, A.Z., and Ghasemi Nejhad, M.N., "Additive manufacturing/3D printing of polymer nanocomposites: structure related multifunctional properties," (Chapter 4, pp. 87–113) *Structure and Properties of Additive Manufactured Polymer Components*, (K. Friedrich, R. Walter, Eds.), Elsevier Publications, Waltham, MA, USA, 2020, ISBN: 9780128195352.

17 A Path Forward for the Composites Industry
Some Final Thoughts

Probir Guha

CONTENTS

We have seen how the use of composites in the form of injection-molded fiber filled thermoplastics, fiber-reinforced reaction injection molded products (RRIM), structural reaction injection molded (SRIM) components, and sheet molding compound (SMC) have been used in automotive applications through the years. However, in the last decade, aluminum and magnesium have gradually created a space in applications targeted for weight reduction. These metals, with their capability to achieve a 40%–50% weight reduction compared to steel, are taking the coveted space of the lightweight, alternative surpassing traditional composites.

SMC not to be undone, successfully introduced a low density SMC as another viable lightweight option. This was attained by replacing a part of the traditional filler, calcium carbonate, with hollow glass spheres. Initially, there was a slight degradation of properties. But a patented treatment of the hollow glass spheres to improve their adhesion to the matrix resin brought back the lost mechanical properties. Even the low-density SMC was unable to put substantial distance between SMC and aluminum in automotive body panel applications and fell short in weight reduction capabilities in structural applications in comparison to aluminum and magnesium.

The industry then turned to carbon fiber. Yet, both material costs and processing costs seemed to be the major hurdle.

A CASE FOR CONTINUOUS FIBER

The industry will need to use continuous glass fibers with a focus on minimizing cycle time or maximizing throughput. Over the years, chopped and short fiber systems in thermosets and thermoplastics have helped in achieving rapid cycle times both in injection molding and compression molding processes; however, over 65% loss in mechanical properties in going from continuous fiber to short fiber has taken away from the weight reduction capabilities of composites. The goal is to focus on continuous fiber without losing the gains made in cycle time over the years.

Process that allows the use of dual fibers with spatial and location flexibility is key – so continuous carbon fibers can be used on a part in regions where the higher stiffness of carbon can be used cost-effectively especially when using carbon fiber processes that are designed to eliminate any fiber wastage without compromising fiber placement and directionality will be most effective.

With companies in the textile industry, now taking an interest in composite, the composite industry now sees how to bring in the advantages of continuous fiber, while eliminating wastage – and without compromising directionality into a rapid cycle process. One such product gaining significant acceptance is Coats Lattice technology.

This path forward of integrating continuous fibers into applications will need to continue for the industry to make the next quantum leap that is required for composites to lay its claim as the most cost-effective weight reduction option.

MARRY STRENGTHS OF THE PAST WITH NEW TECHNOLOGY

To do bigger and better things in the future, we must not forget the original story of composites. In reinventing composites, remembering where it all began is going to be instrumental to continue the tradition of innovation. Our industry will need to hold on to its strengths of short and chopped fiber systems, while bringing in new innovations of continuous directed fibers – while eliminating the weaknesses. How do we maintain the strengths of the past and implement new concepts to create a performance advantage for composites?

In the 1970s, the composite industry tried utilizing preheating the molding compound prior to the heated molding cycle to reduce the mold dwell time required. An effort to improve throughput. This was never fully utilized due to an inability to control the preheating process. This concept would well be revisited for effective improvements in production throughput with updated technology.

Vacuum Molding has been key to the success of SMC in body panels and, indeed, in large molded parts like the truck pickup box. We will need to update and fully understand the engineering of a process like Vacuum Molding to assess how we can scale other molding processes. This determination is especially useful in overmolding processes – indeed, in any process where we can imagine air entrapment causing a deterioration of final product properties.

By definition, composites are comprised of more than one material. The materials of any composite rely on one another to execute the function of the composite. Understanding this, our charge becomes clearer: How can we implement plasma treatment into our processes to maximize the positive interaction between "fiber and resin" and "filler and resin" to further improve the strength and modulus of a molded product? Any increase in efficiency uses less material. By using less material, it becomes a way we can reduce weight...while increasing cost-effectiveness.

COST-EFFECTIVE UTILIZATION OF CONTINUOUS AND CARBON FIBER

The primary strength of composites remains weight reduction capability – but this is at odds with its primary challenge: achieving weight reduction and keeping it cost-effective at the same time. However, both have to be achieved as we bring both continuous fiber and carbon fiber into the product mix.

Commingled fibers and near-net-shape preforms (Coats Lattice technology) from the textile industry is one way of bringing the use of continuous glass and carbon fiber into the realm of mainstream, high-volume composite applications.

- Comolding of glass or carbon fiber in regions of a part that will see higher stress or strain and keeping the rest of the part in the conventional short fiber system is a method that can be used in both thermoset and thermoplastic composite applications. A lot of development will need to take place in the field of over molding for this to come true.
- Using high permeability continuous fiber Lattice preforms allows us to better utilize this rapid cure resin transfer and liquid composite molding processes.
- A third innovation is the application of room temperature stable thermoset chemistry on a reinforcing fiber and then using the coated fiber to make the near-net-shape preforms further improves the viability of this capability. A patented innovation to apply a room temperature thermoset chemistry on a fiber can be expected to enhance the near-net-shape preform technology [1].

If we are able to put a room temperature stable matrix on the continuous fiber prior to making the near-net-shape preform, this technology can be used in conjunction with conventional short fiber molding compounds such as SMC. This may even allow a comeback of injection molded bulk molding compounds (BMC) into structural applications.

- The Lattice technology can be further used to simplify charge placement and transfer during the molding process. This leads to further improvements in molding efficiencies by eliminating "non-value" added steps that are customary of current systems.

DEVELOP DESIGN TOOLS THAT CONSIDER THE DIRECTIONAL NATURE OF COMPOSITES

Materials, process, and design are all interdependent. Alone, they each don't have much value, but together, they strengthen one another.

Traditionally, products designed in composites in automotive have relied on design software and techniques that were often derived from design software used for steel components. This has often led to the current design of chopped fiber systems being based on discounted properties of the composite, to account for property variability caused by fiber flow and ensuing directionality of fiber. This undoubtedly led to over-design and "sub-optimal" design and use of composites. Sure, it's the way "things have always been done"…but how can we improve the way they *can be done* in the future?

A key to unlocking this puzzle lies in developing software specific to composites. Software that integrates composite rheology, reaction kinetics, and product mechanical design in a single piece of design software would go a long way toward addressing the nuances of designing chopped fiber systems. This has not been possible to date. However, knowledge, technology, and methods evolve. To leverage the strides being made in technology, developers require a better understanding and engineering definition of composite material rheology (flow) and thermoset matrix chemical kinetics. They would then need to understand how to integrate this digital understanding into the design of a molded composite. Thus, we could greatly reduce the current state of sub-optimal design of composite products.

The Lattice near-net-shape technology referenced earlier shows how to use continuous fibers very effectively in composite applications. As a first step, this requires the engineer to convert a three-dimensional product shape to a two-dimensional preform shape. The two-dimensional flat shape will need to be outfitted with natural "slits" – allowing it to come together as a three-dimensional shape in a mold. It will be essential to avoid any deformation of fibers during the forming process and to locate the slits in low-stress areas in a manner that avoids creating new premature failure points. A way to address this is to leverage industry tools available for performing the fabric draping simulation. It includes comprehensive fiber simulation to predict the manufacture of components effectively. The user can use different techniques in order to control boundaries, splits, order of drape, etc.; these options allow the users to improve the quality of the final product even before running any stress analysis. There will, however, always be a risk – are we considering the various fiber forms, fabric characteristics that are the basis of this software? Our task as engineers and scientists will be to understand the material, mechanisms, the process, the weaknesses and mitigate these risks.

Increasingly, we will see the use of both the conventional short fibers and the rapidly emerging continuous fibers in the same part. Judicious use of continuous fibers in "high-stress areas." Can this process of design part with both short and continuous fibers be pulled away from the realm of "intelligent decision-making" to being driven by science and engineering that has been captured in design software that considers the special characteristics of short and continuous fiber composites and the performance requirements of the final product?

Materials and process engineers often have to depend on the design engineering practitioners to understand and ratify changes and improvements they may be considering in a material formulation or a molding process. Often in the pursuit of timely completion of a project, we tend not to bother the already overworked design team and indulge in the use of "gut feel," "experience," and "this is how we have done it in the past" scenarios. Sure these are essential ways of getting the job done – and I've even used such subjective instincts in the past, but as our field continues to evolve, is there a way to help increase precision through better software? Perhaps an "easy-to-use" tool based on engineering assumptions would be very useful for materials and process teams to use as they make changes and improvements. So one may raise a question – a wish – is it possible to develop a design tool that can be used for simple and rapid. I know that nothing is "simple" – to ratify material and process changes with respect to product design rapidly?

UNDERSTAND AND UTILIZE SMART COMPOSITES

Integrating smart devices and actuators into a molded composite component is an immense opportunity to add value that is not as easily and cost-effectively added via other technologies like stamping or casting. Our industry will need to master this technology rapidly to provide the market another reason to choose composite products.

Smart composites and actuators that are integrated into the molding operation can be a significant boost to the value-added features of a molded component.

A couple of factors will be key to integrating smart tools into the composite portfolio:

- *Understand the key sensory features* of smart composites and actuators – motion, temperature, facial features, humidity, health attributes, etc.
- *Focus heavily on how these "hardware" can be incorporated* in a molded part. Smart composite hardware and resulting composite piece must be able to withstand both manufacturing process and final use environment without adversely affecting the primary features of the molded composite part.
- *Develop a new practice* to source experts who can infuse this into the industry.
- *Always keep educating future innovators* as this is a rapidly emerging field. The industry expectation is universities will introduce courses and research topics that create a concurrent understanding of composite molding processes and smart devices and actuators.
- *Use data intelligently* to guide decision-making and always question our basic assumptions, backing all hypotheses up with facts.

The truth shall set us free. Never be afraid of true data. Do not manipulate data seemingly to your advantage. Rather, find better, connected, faster ways of continuously learning from data – particularly to take heed of new techniques and methods to innovate and move forward rapidly. We should always keep in our mind the methods and learnings of the past, but unless we are bold enough to embrace new techniques, we won't realize our full potential. When we innovate, we must be willing to "experiment" with the new and that is true of methods and techniques we use in our innovation process as well.

We have cited case studies of how Artificial Intelligence has been used in other areas to collect and analyze data, continuously improve what we learn from the data, and then use it to process and respond to market needs, thereby improving our products. The same can be done for composites. Start small and then scale – but we need to start. In all honesty, there are composite-related activities that are starting to embrace artificial intelligence.

Can this attribute be conjoined with the strengths of smart composites?

Is it possible to gather data using smart devices and actuators that have been cost-effectively integrated into molded parts and use the information in conjunction with other information from "other sources" and analyze and learn and improve performance by using artificial intelligence? We must be open to change – and this is a major change – because if we do not change, we, the composites industry, will miss an opportunity to pull ahead.

SUSTAINABILITY AND ENVIRONMENT FRIENDLY

Another area of innovation lies in the industry's ability to build sustainable solutions. Carbon emissions, weight reduction, and sustainability go hand in hand. A weakness in any of these areas ends up impacting strengths in the other.

In the past, the need to be able to recycle or have recycled content in a composite product was a marketing need. Steel and aluminum were recyclable and were "pulling ahead." We thought that steel and aluminum had a head start and that is why they were ahead in the recycling race.

In today's world, sustainability and a focus on the environment are front and center. This interest and need are societal. Furthermore, competition in this industry will be defined by which companies are leaning into sustainable solutions. Developing sustainable solutions will be an ever-increasing existential challenge for composites.

Increasingly major molders, universities, start-ups are starting to focus in this area. And we have tried to recognize and report on some of those efforts in this book. This is the good news. But the effort needs to dig much deeper to find the correct solution.

All the gains a composite application can make through weight reduction capabilities will be lost if we do not recognize and understand the environmental needs holistically.

A solution will have to consider:

- Life cycle analysis of a component
- Disposition of both post-industrial and end-of-life components
- Total cost of a product starting from raw material to manufacture to end-of-life disposal
- Cost to collect and segregate both post-industrial and end-of-life components based on composites
- Methods and processes to regenerate useful products from post-industrial and end-of-life components based on composites

- And all of this has to be done in a commercially viable manner
- The market wants a safe product at the lowest cost – subsidies are unlikely to sustain a recycle stream
- Collaborating with universities can only make the industry smarter and also attract new talent

INDUSTRY–UNIVERSITY RELATIONSHIPS

It is very simple. Game-changing progress of any kind will require completely new thoughts from outside the field. Failing to recognize the limits of our current work and the benefits of crossover technology will only perpetuate an unending cycle of churning the same "solutions" over and over again. Not having a strong, independent, knowledgeable external input will alter the new entrants into a field back into the "sameness of conventional" ideology. It's a way of thinking that tends to reject new concepts and thoughts and self-inquiry. All essential components to make progress.

It all starts in school. The relentless training, the questioning, the independent thinking. The composite industry must increase its effectiveness in harnessing this strength not only for new ideas or a remote think tank, but to nurture new talent with new thoughts for the future of the industry. Start small but think big in terms of cultivating a relationship with universities. Do not be afraid to think global. Lean into that challenge. Our ecosystem for innovation success is global in nature.

As we bring in new essential attributes into composites – smart composites, data analytics, recycling, etc. – a lot of that requires an open mind and "research" and may best be initiated remotely at a university.

CONCLUSION: KEY FACTORS IN THE PATH FORWARD

In the past several discussions, we have had a chance to review the growth of composites, emergence of alternate technologies, new emerging methods and technologies, etc. We can summarize areas that will help composites reinvent themselves and provide solutions in the future in the following points:

1. Keep the strengths of the past – vacuum molding, short fiber gains in molding efficiencies
2. Continue to improve throughput – molding cycle time reduction
3. Extend plasma and other treatments from bonded surface to improving resin–fiber interface
4. Bridge the gap between materials and process and design techniques to foster more effective product designs
5. Bring in continuous, directed fiber solutions into the high-volume product mainstream
6. Learn from textile solutions for fiber handling solutions and introduce them in composites
7. Incorporate smart composites and artificial intelligence into our solutions

8. The market, the customer knows best – listen to the customer
9. Be open to and promote new thoughts, new ideas, and new solutions
10. Understand how our products affect the environment and find sustainable and cost-effective solutions

REFERENCE

1. Adzima, L.J., Ponn, III, F.H. - Patent - US20040126553- Method for making a charge of moldable material.

Epilogue
The Inspiration – "The Chemistry Does Not Lie"

Probir Guha

CONTENTS

THE EARLY YEARS

Chemistry and honesty intersected for me at an early age.

In my formative years, at home and at Jesuit school, I learned the fundamental importance of truth and honesty. On my father's side of the family, there was a similar premium placed on chemistry. My father, a pharmaceutical scientist, had a favorite saying: "The chemistry does not lie."

My mother, on the other hand, inspired a strong work ethic – not just within me, but my brothers also. "Up before the sun and hit the books," she would say. Getting up that early was painful, but it was worth it. That early training would be instrumental in helping me channel my passion for chemistry into industry-defining innovations. All those individuals who I had the honor of working with – some of the coauthors in this labor of love – will testify to the fact that Probir gets started early, very early in the wee hours of the morning! I believe this led to the foundation on which I built my love for a career in the research of polymers and chemistry.

Yet, therein lies one more lesson my parents taught me: Respect. My mother would make it her mission to ensure her three sons learned the values of respecting others – and letting that define their work.

These lessons, in a way, are the DNA not only of this work but also of my life's work. For those very important early days, I have to respectfully remember and thank my parents, Dr. Jahnavi Ratan and Rekha Guha. My commitment to working hard and always striving to do better is one way I can honor their memories.

Outside of my family home, however, I found another voice to guide my pursuits – Father Camille Bouche, Prefect of the school and a singular Jesuit priest who influenced the lives of countless students who graduated from St. Xavier's School in Kolkata, India.

Father Bouche instilled in us the unarguable merits of punctuality and – yes, there it is again – honesty. My coauthors herein know that Probir is never late to meetings. It's just a way that one of those key lessons continues to drive me.

Life is not just about work – but about the balance and the value that experiences outside work can bring. This effort is very personal and encompasses a lifetime of experiences that has evolved into an extremely satisfying career in fiber-reinforced composites. As I sat down to share some of my thoughts that took me through this journey, I began recalling varied experiences in my lifetime were key to my incredible journey.

Following my days in St. Xavier's, I was fortunate to be chosen to attend the Indian Institute of Technology in Kharagpur, a quaint little university town, about a 2-hour train ride from our home in Kolkata. It was tough to get accepted in the IITs – there were five IITs around India in those days. The institutes were built and funded by the Indian government in the pursuit of fostering a stronger and more independent India.

At IIT, we received the best engineering education in the country. The institute attracted the finest students the Indian school system had to offer – and that's why I have always felt that the five years I spent at IIT allowed me to grow considerably, as I interacted with the best minds of the country – and perhaps even of the world. I have to collectively thank my IIT friends for inspiring me and pushing me to be the best – you know who you are.

Education, however, is a lifelong journey and it didn't just end after IIT. So, then I came the United States and I was introduced to the world of polymers. At the University of Detroit, many interactions influenced me, but none more than the guidance from one of the leading figures in automotive composites, Dr. Joseph Epel – "Doc" as he was called by those of us who knew him – was a brilliant teacher, chemist, researcher, and entrepreneur. Doc taught me that, without a well-thought-out and defined objective, any effort would fail. This is a very simple concept but so important to understand and practice. Doc taught me, influenced me, and gave me opportunities; he encouraged me to learn from my failures before I could truly succeed. His influence on my career in composites remains very significant to this day – and I am forever in debt to Doc for taking me under his wing and sharing his wisdom – as I was learning to navigate a new world and a new career.

As I found my place in the world of composites and advanced in my career, I took note of the influence that people outside of my day-to-day could have on my work. Very often in any endeavor, the people who contribute the most influence to the outcome might not be the people I see at work every day; they may not even live within the world of automotive composites as we do. But it's from them that I drew – and continue to draw – strength.

Through the years, my wife Sriradha has been a pillar, championing the same virtues with which I was raised – honesty and a strong work ethic – and providing a north star for those moments of doubt. As I made the leap from India to the United States and beyond, she has always been my copilot, infusing cool-headed, but clever perspectives – and always steering me back on course.

Perhaps the most unexpected – but not unsurprising – source of support comes from my two sons, Rahul and Rohin. They have often unknowingly contributed, motivated, and taught me in more ways than they will ever know. Sometimes when the chips were down I learned from their life experiences how I could overcome what appeared to be an insurmountable problem.

And a special thanks to Rohin, a wizard at putting thoughts into words and our family wordsmith, for his help with the "Guha Chapters" in this compilation. I thank both of them for their support through the years and for motivating me to take on this project.

Support and inspiration run in the family – and I want to thank my brothers Pronob and Prateep who, like me, also paid close attention to the lessons of honesty and hard work our parents taught us. Pronob, however, paved the path in the early days – being the first one to enter St. Xavier's and the IIT and the other two followed in his footsteps. In the days leading up to this publication, my discussions with Prateep on how best to harness 40 years of composites experience into a collection of chapters in this book were critically important. His support – whether it's about the importance of data and analytics or the ways we can harness artificial intelligence to keep pioneering across industries – has been essential to the narratives contained in this book.

Everyone in my family has, in their own way, strengthened my core beliefs in truth, honesty, and respect for others. They are fundamental core beliefs and it is this sustained strong foundation that is essential to succeed in more conventional and transparent areas. These are also the virtues I believe will be essential to shaping the future of the world of composites.

A FINAL THANK YOU

My journey within the world of composites has been incredible. Despite being a participant in this world for more than 40 years, it honestly feels like I'm only getting started.

I see new players – like Coats – going all-in into this arena and I'm encouraged about the future of the industry. Seeing outside developers pivot to composites and pioneering new solutions ensures that the tradition of innovation, central to composites, continues.

Many of the talented people I met along my journey have contributed to this effort. This story, however, will continue. With good reason, too. There are so many, (especially) engineers and managers, who are extremely adept at introducing new technologies and products into manufacturing plants, who could fill the pages that follow. The interface between development and manufacturing, bringing the customer into the process of the new solutions and finally making it work in the factory needs talent, knowledge, leadership, and mental fortitude. There are many people instrumental to this type of effort who I have had the privilege to work alongside since my early days in the SMC world.

The right team is critical to bring any innovation to life – and it's for that reason I would like to thank Mayur Shah, who I've known since our days at The Budd Company and eventually to CSP-Teijin. I'm aware that his knowledge, his attention to detail, and his on-the-ground leadership shepherded some of the most successful product launches in the industry, including TCA body panels and the pick-up box.

Similarly, I was lucky to find other catalysts for change in my new environment at Coats. Transformation is never easy, but when you reach the finish line, the achievement feels incredible and I feel honored to have been part of a journey through innovation with great minds like these.

The list of incredible minds who have inspired me is long – and that's because there's no limit to where good ideas come from. Innovation has no single source. In fact, innovation is the common thread running through each exploration in this effort. No matter how varied the contributors in this effort, what they all share is the same passion for engineering solutions across the world of composites. With that in mind, I'm humbled to have featured contributors from a variety of backgrounds: From the user of composites to development engineers to individuals from the customer–supplier interface to individuals from educational institutions who have interacted with the composites industry and even individuals with considerable success in seemingly unrelated industries around the world. The veterans of today and the leaders of tomorrow alike are equally important to the future of composites. Future innovation in this field will be driven by diversity in views and our ability to learn from the past as we embrace new ideas to bring solutions for the challenges of the future.

CONCLUSION

I have had the honor of working shoulder-to-shoulder with a varied group of individuals through the years. Many with whom I have shared this exciting ride most wholeheartedly collaborated in this endeavor. For the fun journey – and for them sharing their insights in this book – I thank them.

We tried to capture all the key thoughts on how we can make this world better through the use of composites. However, this is only part of a larger conversation. I am sure we have defined a platform from where we can create a new trajectory for sustainable growth for composite applications. I want to invite others into this conversation, to fill in the blanks, to introduce new ideas, and to share new discoveries. It's all consistent with the tradition of innovation that defines our industry.

Whenever I needed guidance throughout my journey in composites, I always turned to the fundamentals I was empowered with by my parents and by Father Camille Bouche.

And as I grew up and moved away from home, my wife Sriradha continued to help provide that foundation – her resolve, strength, and support all inspiring me to continue to push through moments of extreme doubt.

The whole story was built on honesty, discipline, hard work, and respect for the fellow human being.

Index

Note: **Bold** page numbers refer to tables and *italic* page numbers refer to figures.

Printed in the United States
by Baker & Taylor Publisher Services